Anaerobic Fungi

MYCOLOGY SERIES

Editor

Paul A. Lemke

Department of Botany and Microbiology
Auburn University
Auburn, Alabama

Additional Volumes in Preparation

Anaerobic Fungi

Biology, Ecology, and Function

edited by

Douglas O. Mountfort

Cawthron Institute
Nelson, New Zealand

Colin G. Orpin

Division of Tropical Crops and Pastures
CSIRO
Brisbane, Queensland, Australia

CRC Press
Taylor & Francis Group
Boca Raton   London   New York

CRC Press is an imprint of the
Taylor & Francis Group, an **informa** business

First published 1994 by Marcel Dekker, Inc.

Published 2019 by CRC Press
Taylor & Francis Group
6000 Broken Sound Parkway NW, Suite 300
Boca Raton, FL 33487-2742

© 1994 by Taylor & Francis Group, LLC
CRC Press is an imprint of Taylor & Francis Group, an Informa business

First issued in paperback 2019

No claim to original U.S. Government works

ISBN-13: 978-0-367-44944-5 (pbk)
ISBN-13: 978-0-8247-8948-0 (hbk)

Visit the Taylor & Francis Web site at
http://www.taylorandfrancis.com

and the CRC Press Web site at
http://www.crcpress.com

Library of Congress Cataloging-in-Publication Data

Anaerobic fungi : biology, ecology, and function / edited by Douglas O. Mountfort,
Colin G. Orpin.
     p.   cm. — (Mycology series; v. 12)
  Includes bibliographical references and index.
  ISBN 0-8247-8948-2 (acid-free paper)
  1. Anaerobic fungi. I. Mountfort, Douglas O. II. Orpin, Colin G. III. Series.
QK604.2.A53A53 1994
589.2'04128—dc20   94-19072
                                           CIP

Series Introduction

Mycology is the study of fungi, that vast assemblage of microorganisms which includes such things as molds, yeasts, and mushrooms. All of us in one way or another are influenced by fungi. Think of it for a moment — the good life without penicillin or a fine wine. Consider further the importance of fungi in the decomposition of wastes and the potential hazards of fungi as pathogens to plants and to humans. Yet, fungi are ubiquitous and important.

Mycologists study fungi either in nature or in the laboratory and at different experimental levels ranging from descriptive to molecular and from basic to applied. Since there are so many fungi and so many ways to study them, mycologists often find it difficult to communicate their results even to other mycologists, much less to other scientists or to society in general.

This series establishes a niche for publication of works dealing with all aspects of mycology. It is not intended to set the fungi apart, but rather to emphasize the study of fungi and of fungal processes as they relate to mankind and to science in general. Such a series of books is long overdue. It is broadly conceived as to scope, and should include textbooks and manuals as well as original and scholarly research works and monographs.

The scope of the series will be defined by, and hopefully will help define, progress in mycology.

PAUL A. LEMKE

Preface

During the past fifty years our knowledge of microorganisms in the gut has developed to the extent that the microbial ecology of such ecosystems, particularly the rumen, are some of the better understood. However, although many of the important gut bacteria and protozoa have been cultured and their metabolism well characterized, only recently has it become apparent that a third group of organisms, the anaerobic fungi, represent an important part of the gut microflora.

This volume brings together for the first time information on anaerobic fungi considered from a number of viewpoints. It is intended to provide a comprehensive account of their taxonomy, physiology, and ecology. The nucleic acids of these organisms are described together with gene cloning, ribosomal RNA sequence, and establishment of molecular phylogeny. Methods for the enumeration of fungi in gut environments are also detailed. The work deals mainly with fungi from the rumen but also gives account of fungi from other gut environments, such as the cecum and hindgut of nonruminant herbivores.

The work should be considered as a supplement to previous monographs on the microbial ecology of the gut. These include the monographs of Hungate, *The Rumen and its Microbes* (1966), Clarke and Bauchop, *The Microbial Ecology of the Gut* (1977), and Hobson, *The Rumen Microbial Eco-*

system (1988). The text of Draser and Hill, *Human Intestinal Flora* (1974), may also serve as a useful background reference together with Zehnder's *Biology of Anaerobic Microorganisms* (1988) with its emphasis on the principles of anaerobic microbiology.

This volume will serve as an important reference for researchers in the field of anaerobic microbiology and, in particular, the microbiology of gut ecosystems. We also hope it serves as a useful reference to students of general microbiology, and as a text for those specializing in gut microbiology.

In conclusion, we wish to thank the contributors for their cooperation, and other colleagues, who have stimulated our interest in this field.

DOUGLAS O. MOUNTFORT
COLIN G. ORPIN

Contents

Contributors

Danny E. Akin, Ph.D.   Research Microbiologist, Richard B. Russell Agricultural Research Center, Agricultural Research Service, U.S. Department of Agriculture, Athens, Georgia

Alan G. Brownlee, B.Sc.Hons., Ph.D.   Principal Research Scientist, Division of Animal Production, CSIRO, Prospect, New South Wales, Australia

David R. Davies, B.Sc., Ph.D.   Nutrition and Survival of Anaerobic Fungi, Microbiology Group, Department of Environmental Biology, Institute of Grassland and Environmental Research, Plas Gogerddan, Aberystwyth, Dyfed, Wales

G. Fonty   Laboratoire de Microbiologie, Institut National de la Recherche Agronomique, Saint Genes Champanelle, France

James France, D.Sc., C.Math.   Group Leader, Biomathematics Group, AFRC Institute of Grassland and Environmental Research, Okehampton, England

E. Grenet   Institut National de la Recherche Agronomique, Saint Genes Champanelle, France

Keith N. Joblin, Ph.D.   Department of Rumen Microbiology, AgResearch, Grasslands Research Centre, Palmerston North, New Zealand

Douglas O. Mountfort, Ph.D.   Scientific Programme Leader, Marine Microbial Enzymes and Antarctic Anaerobes, Cawthron Institute, Nelson, New Zealand

Edward A. Munn, B.Sc., Ph.D., D.Sc., C.Biol., F.I.Biol., Dip.R.M.S.   Department of Immunology, The Babraham Institute, Babraham, Cambridge, England

Colin G. Orpin, B.Sc., Ph.D., C.Biol.*   Program Manager, Applications for Biotechnology, Division of Tropical Crops and Pastures, CSIRO, Brisbane, Queensland, Australia

Michael K. Theodorou, Ph.D., B.Sc.   Group Leader, Microbiology Group, Institute of Grassland and Environmental Research, Plas Gogerddan, Aberystwyth, Dyfed, Wales

Alan G. Williams, D.Sc., Ph.D.   Department of Food Science and Technology, Hannah Research Institute, Ayr, Scotland

Nigel Yarlett, B.Sc., Ph.D.   Associate Professor, Department of Biology, and Research Associate, Haskins Laboratories, Pace University, New York, New York

---

*Present affiliation: Consultant in Agricultural Biotechnology, Cambridge, England

Anaerobic Fungi: Taxonomy, Biology, and Distribution in Nature

COLIN G. ORPIN*

Division of Tropical Crops and Pastures
CSIRO
Brisbane, Queensland, Australia

I. THE DISCOVERY OF ANAEROBIC FUNGI

Herbivorous mammals depend for their survival on a symbiotic association with microorganisms in their alimentary tract. The diet of grazing and browsing herbivores consists largely of plant structural carbohydrates, such as cellulose and hemicelluloses, that the animals themselves are unable to digest. Instead, the symbiotic microorganisms in the alimentary tract, particularly in the rumen of ruminants and the cecum and hindgut of nonruminants, hydrolyze these compounds under anaerobic conditions, with the production of microbial cells and volatile fatty acids that the animal can utilize as sources of nutrition (1,2,3). In order to understand and control digestion of plant structural carbohydrates and improve the production efficiency of ruminants, the microbial population of the rumen of domestic sheep and cattle has been studied extensively (1,2,3).

The microbial population of the rumen is diverse, and until the discovery of anaerobic fungi, it was believed to consist principally of anaerobic and facultatively anaerobic bacteria, ciliated protozoa, and flagellate protozoa. Early workers (4,5) documented the existence of uniflagellate, biflagellate, and multiflagellate organisms in rumen contents, believing them to be flagellate protozoa. These organisms were placed in the genera *Callimastix, Oikomonas, Monas,* and *Sphaeromonas.* Multiflagellate organisms similar

*Present affiliation: Consultant in Agricultural Biotechnology, Cambridge, England

to the *Callimastix frontalis* described by Braune (5) in the rumen were subsequently found in different habitats; *Callimastix equi* Hsuing was found in the horse cecum (6), *Callimastix jolepsi* in the pulmonate snail (7), and *Callimastix cyclopis* in the copepod *Cyclops stenuus* (8). The status of the multiflagellate organisms from the horse cecum and the pulmonate snail remains to be determined, but *Callimastix cyclopis* has been examined in some detail (8). It was found that the flagellate was in fact a zoospore of a fungus with a plasmodial vegetative stage that developed in the body cavity of the host copepod and, on maturity, gave rise to the flagellates. It was believed that the flagellates would then infect a new host to continue the life cycle. Since this species was identified as a fungus and not a protozoan, it was proposed that the rumen flagellates with multiple flagella, which were still assumed to be flagellate protozoa, should be reclassified as a species of zooflagellate in the new genus *Neocallimastix* (8), with *frontalis* (5) as the type species.

Many of the flagellated rumen organisms described by Liebetanz (4) and Braune (5) have since been shown to be not protozoa, but the flagellated zoospores of anaerobic fungi (Fig. 1). The first report of isolation of an anaerobic fungus, a species of *Neocallimastix*, was published in 1975 (9). The organism was isolated during attempts to isolate and culture anaerobic flagellate protozoa from the rumen contents of sheep using a published procedure (10). Flagellates did indeed grow in the cultures, but it was impossible to separate the flagellates from what appeared to be vegetative fungal growth in the culture. It was soon evident that the flagellates were released from reproductive structures borne on the fungal rhizoids, and that the life cycle of the organism consisted of an alternation between a motile, flagellated zoospore stage and a vegetative rhizoid-bearing reproductive stage. The organism was similar both in morphology and life cycle to a chytridiomycete fungus, but it was a strict anaerobe. Until then, fungi were regarded as being either anaerobes or facultative anaerobes, and the detection of microorganisms similar to chytridiomycete fungi that could grow only under chemically reducing conditions in the absence of molecular oxygen was novel. Because of the revolutionary nature of this finding, acceptance by the scientific community was slow. The presence of chitin in the cell walls of these and similar organisms (11) confirmed that they were true fungi despite being strict anaerobes.

Methods for the isolation and culture of anaerobic fungi have since been considerably refined, so that the organisms can now be routinely isolated from suitable habitats with little difficulty. Anaerobic fungi are now regarded as a normal component of the microbial population of the rumen.

The reason why the rumen fungi remained undiscovered while research on rumen bacteria and protozoa went ahead strongly during the period up to 1975 is not difficult to understand, even though electron microscopy was commonly employed to determine microbial activity during the digestion of

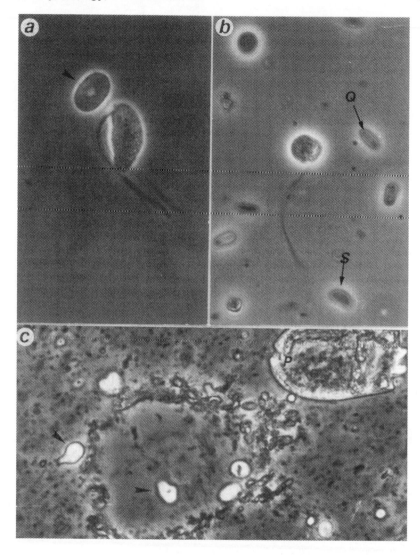

Figure 1 Flagellate zoospores of anaerobic fungi as observed in fresh rumen contents. **(a)** A multiflagellate zoospore, possibly of *Neocallimastix* sp., is shown, together with the large bacterium *Magnoovum eadii* (arrowed). **(b)** A uniflagellate zoospore is shown, together with Quin's Oval (Q) and another large bacterium — *Selenomonas ruminantium* (S). The zoospores can be clearly differentiated from the large bacteria. **(c)** Uniflagellated zoospores (arrowed) can be seen in relation to a ciliated protozoan, *Entodinium* sp. (P). Phase-contrast microscopy. **(a)** × 2000; **(b)** × 2000; **(c)** × 700.

plant tissues in the rumen and the flagellates of the rumen fungi had been described (as flagellate protozoa) as early as 1910 (4). Work on the flagellate protozoa of the rumen was limited because of their small population densities and the consequent assumption that they were of little importance in rumen metabolism; hence the fungal flagellates were also largely ignored. In addition, it was common for rumen microbiologists to strain rumen contents through cheesecloth to remove the large plant fragments before microbiological analysis, as pointed out by Bauchop (12). In doing this, the microbiologists were separating the bulk of the vegetative growth of the fungi from their working material, and it was only after the introductory work of Orpin (13,14) on the invasion of plant tissues by the flagellates of the rumen fungi, and the scanning electron microscopy of Bauchop (15), that the significance of separating the plant particles from rumen contents was recognized in this context. The fungal growth is normally tightly attached to digesta fragments (Fig. 2). However, rough handling during the straining of digesta through muslin sometimes damages the vegetative fungal growth and breaks sporangia from the rhizoids (Fig. 2b). These sporangia can be differentiated from protozoa (Fig. 1c) in the filtered rumen fluid by their lack of motility, high refractivity, and lack of cilia; they contain no skeletal plates and usually have a short length of rhizoid attached.

Additional reasons why the rumen fungi remained undiscovered until recent years were the difficulty of isolating them from rumen contents without the use of antibiotics to suppress the growth of bacteria (9,14,16,17) and the need to isolate them from low dilutions of rumen contents. This is usually within the range of 10^{-3}–10^{-5}, well below that normally employed for the isolation of anaerobic rumen bacteria. Thus colonies of anaerobic rumen fungi would probably not have been observed during the isolation of rumen bacteria.

The importance of these fungi in herbivore nutrition is still not well understood, but their undoubted ability to utilize major plant cell wall polysaccharides for growth (18–20) and to produce a wide range of enzymes with activities capable of hydrolyzing many of the components of plant cell-walls (21,22,23) shows that they have the potential to contribute substantially to plant-fiber degradation in the alimentary tract of the host animal.

The life cycles of all anaerobic fungi so far described consist of an alternation between a motile, flagellate zoospore stage, free-living in the liquid phase of the digesta, and a nonmotile, vegetative, reproductive stage, saprophytic on digesta fragments in the alimentary tract of the animal. The zoospores of all species are able to change to an ameboid form and may be observed crawling over digesta fragments and, in culture in vitro, the sides of culture vessels. Evidence is accumulating that in some species (isolated from

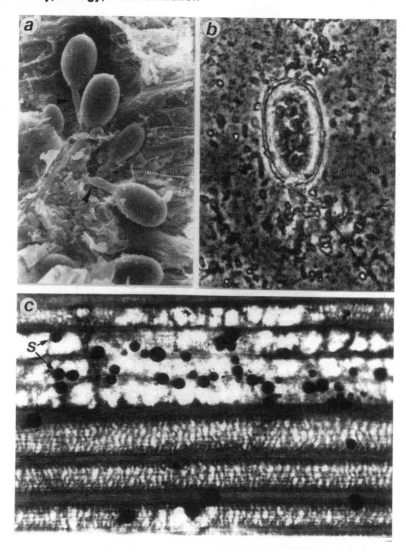

Figure 2 Vegetative stages of anaerobic fungi found in rumen contents. In the scanning electron microscope (a), the sporangia of the fungi can be seen with the associated rhizoid (arrowed) penetrating the plant tissue substratum. The sporangia (S) are broken from the plant tissues (b) by rumination and straining of rumen contents. Plant tissue invasion can be extensive (c) where zones of dissolution of plant tissue around the fungi (S) are evident. Phase-contrast microscopy was used in (b) and (c). (a) ×210; (b) ×460; (c) ×100.

ruminants) an additional oxygen-tolerant stage may be developed in the feces. An oxygen-resistant stage, possibly a sporangium, is suspected (24), but in an isolate from the horse cecum, a possible resting stage unlike a sporangium has been identified (25). In the coelomic parasite *Coelomyces* (= *Callimastix*) *cyclopis* of *Cyclops stenuus*, a resting cyst gives rise to biflagellate zoospores that are the infective stage for the mosquito, a second host of the parasite (26); as yet no such cysts have been observed in the anaerobic fungi, unless the putative resting stage of an isolate from the horse cecum (25) is equivalent. A schematic, generalized life cycle of anaerobic fungi from the rumen as we understand it at present is shown in Fig. 3.

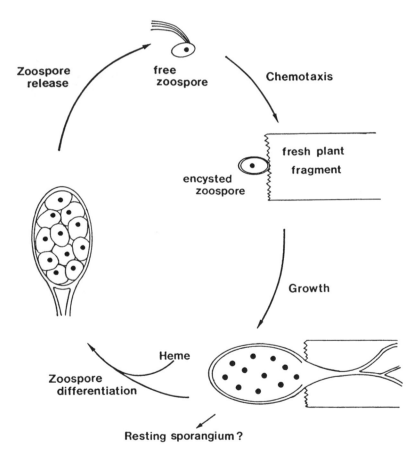

Figure 3   Schematic life cycle of anaerobic rumen fungi.

There is much variety in the flagellation of the zoospores from species to species, in the morphology and the development of the rhizoidal system, and in the mode of growth of the genera of anaerobic fungi so far characterized. Some species have either predominantly uniflagellate or multiflagellate zoospores; some species are monocentric, with a single sporangium on each thallus, mostly with endogenous development of the zoospore cyst to produce the sporangium. Others are polycentric, with exogenous development of the zoospore cyst to produce a multinucleate rhizomycelial system eventually bearing multiple sporangiophores that expand at their tips to produce the sporangia. The rhizoidal system may be simple, composed mostly of spherical structures when grown in vitro, or fibrillar and extensively ramifying. Consequently they do not fit comfortably within any previously existing taxon.

All of the anaerobic fungi so far isolated live in the rumen and reticulum of ruminants, the forestomach of camelids and macropod marsupials, or the cecum and large intestine of other, mostly large, herbivores. In ruminants they can be isolated from all regions of the alimentary tract and from the feces, but there is little evidence to suggest that they actually grow in any organ except the rumen. Numerous attempts to isolate them from other habitats, such as anaerobic muds, have been unsuccessful (27,28).

Virtually all our knowledge of these strictly anaerobic fungi has accumulated since 1975; the discovery of such an unusual group of eucaryotes is a rare phenomenon, and this is now reflected in substantial efforts by scientists worldwide to understand their ecology, biochemistry, phylogeny, and role in herbivore nutrition.

II. TAXONOMY OF ANAEROBIC FUNGI

A. Systematics

All known anaerobic fungi are zoosporic, and in modern fungal classifications zoosporic fungi are placed in the subdivision Mastigomycotina of the division Eumycota (29). The Mastigomycotina contains an assemblage of phylogenetically unrelated classes (30) united principally by the production of zoospores, and includes the class Chytridiomycetes. Within the Chytridiomycetes is the order Spizellomycetales (31), in which the anaerobic fungi were originally placed by virtue of the ultrastructural characteristics of their zoospores (32). The family Neocallimasticaceae was erected to include, initially, the multiflagellate monocentric genus of anaerobic fungus *Neocallimastix* Vavra and Joyon ex Heath, but the family description was later emended to include all monocentric, polycentric, uni-, bi-, and multiflagellated anaerobic zoosporic fungi (33).

According to Barr (30), the main problem in the systematics of the Chy-tridiomycetes is that many species display extensive morphological variation. Indeed, after some species were studied in pure culture, the morphological variation was found to be so great as to render of little value many of the specific and generic criteria found in the classical literature. Consequently, transmission electron microscopy was used to confirm chytridiomycete tax-onomy (32). Particular attention has been paid to the fine structure of the kinetosomes and their accessory structures as a tool to characterize the Chy-tridiomycetes. However, morphological development as observed in the light microscope is of considerable value in determining taxa, provided the strains being examined are grown in defined, constant media (33).

In the anaerobic fungi, as in other chytridiomycetes, there is great varia-tion in the morphology of pure isolates grown in different media and within a single culture at different stages of maturity. This is particularly true for the *Neocallimastix* spp. and *Piromyces* spp. For this reason it is recommended that in studies of comparative morphology and the identification of new isolates, the medium of Heath (34) be used if possible.

It is clear that the various genera of the Neocallimasticaceae that have been examined at the fine-structural level (32,35–39) show a remarkable degree of conservation in many features and may represent a distinct taxon with some dissimilarities from the Spizellomycetales, such as in the process of mitosis (40) and the universal distribution of hydrogenosomes in the anaer-obes. The description of Spizellomycetales Barr (31) was emended (32) to accommodate the Neocallimasticaceae.

The taxonomy of the gut fungi is still under extensive examination, and Heath and his co-workers have recently proposed a new order, Neocallimas-ticales, within the phylum Chytridiomycota, containing three major clades: the gut fungi, the Blastocladiales, and a Spizellomycetales–Chytridiales–Monoblepharidales complex, based on an extensive cladistic analysis of struc-tural characteristics (41) and rRNA sequences (42).

Neocallimasticaceae Heath (32) emended Barr et al. (33).

Monocentric and polycentric, uniflagellate, biflagellate, and multiflagellate obligate anaerobes in the digestive tract of vertebrates.

Although the family description as it exists now allows for the occurrence of members with zoospores carrying two flagella, to date no species has been described that consistently produces zoospores that are predominantly bi-flagellate. Bi-, tri-, and quadriflagellate zoospores have been found only in relatively small numbers in pure cultures of normally uniflagellate species, such as *Piromyces* (33,43) and *Sphaeromonas* (16). Biflagellate cells, appar-ently zoospores of anaerobic fungi, may be observed in rumen contents (C. G. Orpin, personal observations) and were described by Liebetanz (4). In the rumen, and in cultures in vitro, undifferentiated flagellates may be released from sporangia as multiflagellate structures (9). Bi-, tri-, and quadriflagellate zoospores may represent similar but less extreme forms of undifferentiated zoospores of uniflagellate species.

B. Monocentric genera and species

Neocallimastix Vavra et Joyon ex Orpin and Munn, emended (44).

Zoospores spherical to broadly ellipsoidal; organelles sometimes clustered in the posterior half; nucleus equatorial or central; 9–17 flagella inserted mainly in two rows; plasmalemma coated with elaborately organized fibrous material.

The genus *Neocallimastix* was erected for (monocentric) species with multiflagellate zoospores, with *frontalis* Braune as type species (13), the type species being isolated from a cow. The genus description was emended by Orpin and Munn (44) to include *patriciarum* (Fig. 4), isolated from sheep. There is some confusion in the literature concerning the identity of the organisms used in some of the early work. Early work of Orpin's group was conducted with a strain of *Neocallimastix* that was originally designated *N. frontalis* following Braune (5). When *Neocallimastix frontalis* was formally named by Bauchop and his co-workers (32), it was clear from the species description that the *frontalis* of Orpin's group was not the same as the *frontalis* of Bauchop's group, and the former strain was subsequently described as *N. patriciarum* (44). Thus, early papers of Orpin's group, before 1986, refer in fact to *N. patriciarum* and not *frontalis*.

Photomicrographs of *N. patriciarum*, representative of the genus, are shown in Figs. 4 and 5. The rhizoidal system of the genus is shown in Figs. 4 and 5. The rhizoid system (Fig. 4a) of the vegetative stage may extend to 2.5 mm in soft agar culture. The zoospores are shown from a living preparation (Fig. 4b), in which the flagellar bundle is held tightly together in a single locomotory organelle, and from a formaldehyde-fixed preparation (Fig. 4c), in which the individual flagella within the locomotory organelle can be clearly seen. Ameboid movement of a zoospore of *N. patriciarum* is shown in Fig. 4d, with the flagella trailing posteriorly. Release of zoospores from the sporangium of *N. patriciarum* is shown in Fig. 4e; here the zoospores are fully formed and are spinning rapidly within the partially dissolved wall of the sporangium. No evidence for a defined papilla, through which the zoospores might be released, can be seen. Attachment of zoospores to solid surfaces before encystment is mediated by the production of fibrous material (Fig. 5a). Vegetative growth of *N. patriciarum* growing on perennial ryegrass in vitro is shown in Fig. 5b, with rhizoids associated with the surface of the plant tissue as well as penetrating the tissue. The major sites of invasion and growth of the fungi appear to be the stomata and abraded and cut surfaces.

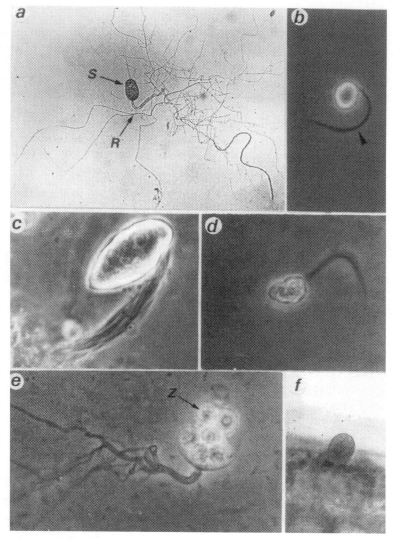

Figure 4 *Neocallimastix patriciarum.* The vegetative stage **(a)** is monocentric, consisting of a single sporangium (S) borne on a much-branched rhizoidal system (R). The living zoospore **(b)** appears to have a single locomotory organelle (arrowed), which in fixed preparations is seen to consist of 12–17 flagella **(c)**. Zoospores can crawl by ameboid movement **(d)** with the flagella trailing. Release of zoospores (Z), here seen spinning within the partly autolysed sporangium **(e)**, invade fresh plant tissues **(f)**, where they encyst and germinate; here at a stoma. **(a)** ×120; **(b)** ×600; **(c)** ×1800; **(d)** ×600; **(e)** ×260; **(f)** ×600.

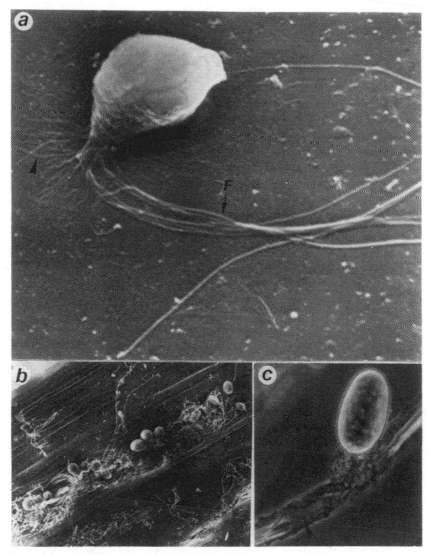

Figure 5 **(a)** Zoospore of *Neocallimastix patriciarum* attached to a solid substratum by unidentified fibrillar material (arrowed), with some flagella (F) still attached. **(b)** When grown in vitro. *N. patriciarum* develop rhizoids (R) that spread over the tissue surface as well as penetrating the tissue. **(c)** Growth of the fungus on the awn of barley (c) clearly demonstrates the removal of tissue components (arrow) revealed by loss of refractivity. **(a)** ×4000; **(b)** ×60; **(c)** ×320.

N. frontalis (Fig. 6) differs from *patriciarum* in having a less extensive rhizoidal system, a bipolar distribution of intracellular organelles, and differences in fermentation product pattern.

Another species of *Neocallimastix, N. hurleyensis* (39,45), has been described from cattle. It was originally designated strain R1 (20) in earlier publications, and it differs from both *frontalis* and *patriciarum* in that it possesses an annular hydrogenosome. Another sheep isolate, clearly a species of *Neocallimastix*, from Stanthorpe, Queensland, Australia (C. G. Orpin, unpublished) has constant major morphological differences from all other published species in that the rhizoids show little bifurcation and are strongly constricted near the base of the approximately spherical sporangium, and the zoospores constantly possess 20–25 flagella. The fine structure of three *Neocallimastix* spp. has been described: *N. frontalis* (32), *N. patriciarum*, (35,37,46,47), and *N. hurleyensis* (45).

The number of flagella possessed by a species of *Neocallimastix* is not a constant criterion of that species, as flagella numbers vary between zoospores even in clone cultures (32,35). It is thus not safe to use the number of flagella alone to differentiate between species unless the new isolate carries a number of flagella that is well outside the limits found within clone cultures. This certainly is true for strain St05, which has 20–25 flagella, compared with the 7-17 possessed by the three previously described species.

Colonies of *Neocallimastix* spp. grown in agar-containing medium for 48 h in vitro consist of a central core of sporangia surrounded by radially developing rhizoids (Fig. 7a), larger than those of *Piromyces* spp. and distinctly different from *Sphaeromonas* (Fig. 6d), which lacks radiating fibrillar rhizoids. This type of colony growth is also distinct from that of the polycentric species, colonies of which do not produce a clearly visible central core of sporangia and consist mostly of intertwined rhizoids (Fig. 6e).

A polycentric, multiflagellate species was isolated from sheep in France and named *Neocallimastix joyonii* (48). It was placed in the genus *Neocallimastix* on the basis of the flagellation of the zoospore. Another polycentric species with multiflagellate zoospores has been described from cattle in Canada (33) for which the genus *Orpinomyces* was erected. Because of the similarity between the vegetative stage of *N. joyonii* and that of *Orpinomyces*, the dissimilarity of the vegetative morphology from that of *N. frontalis* (the type species for the genus *Neocallimastix*), and exogenous development of *joyonii* compared with the endogenous development of *Neocallimastix* spp., the author considers that *joyonii* should now be placed in *Orpinomyces*.

Caecomyces Gold (36).

Zoospore spherical to ellipsoid to ameboid, posteriorly uniflagellate; organelles dispersed throughout the cell or clustered among flagellar roots; microtubular roots

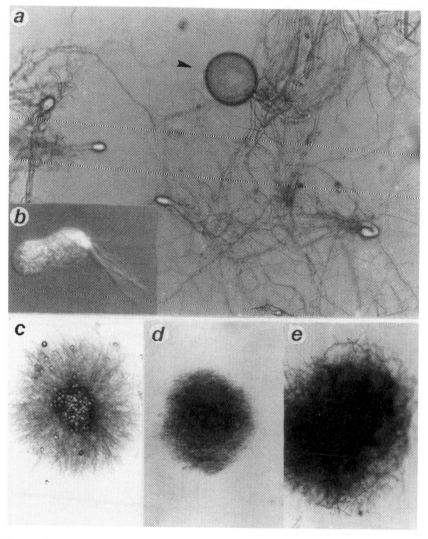

Figure 6 *Neocallimastix frontalis* (**a**) is similar to *N. patriciarum* but has a less extensive rhizoidal system. The mostly spherical sporangia (**a**) (arrowed) and waisted zoospore (**b**) are characteristic of the species. Colonies of anaerobic fungi in agar have radiating rhizoids (**c**), as shown here in *Neocallimastix* spp; or are granular (**d**), as in *Sphaeromona communis*, or dense with matted rhizoids as in *Orpinomyces* spp. (**e**). The colony in (**c**) is 2 d old; colony (**d**) and (**e**) are 3 d old. Phase-contrast microscopy. (**a**) ×200; (**b**) ×1250; (**c**) ×70; (**d**) ×70; (**e**) ×70.

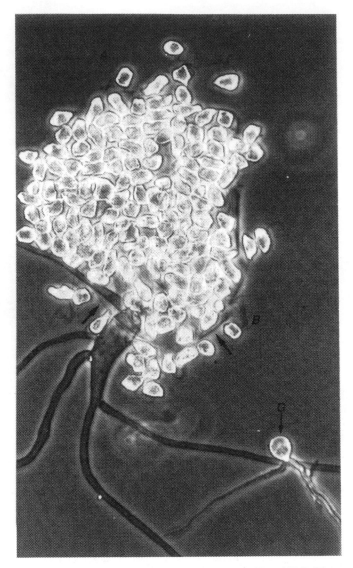

Figure 7   Release of zoospores from *Neocallimastix* St05 at 30°C. The sporangium wall in region A has dissolved, leaving some fragments (arrows) remaining in region B. Because of the low temperature the zoospores have moved little from the dissolving sporangium. A freshly germinated zoospore is shown (G). Phase-contrast microscopy. ×530.

radiate from the kinetosome-associated spur in two arrays, a divergent anteriorly directed array and a fan-shaped array running toward and along one side of the posterior of the cell; flagellar transition zone surrounded by a circumpolar ring connected via two or more struts to a skirt and scoop around the base of the kinetosome; nucleus associated with anteriorly directed roots, usually variously pronounced beak extended to base of kinetosome; microbodies (hydrogenosomes) interspersed among flagellar roots; ribosomes segregated into helices and numerous larger clusters in zoospores; plasmalemma coated with fibrous material; rhizoids predominantly form one or more large bulbous structures with rare attached fibrillar roots, or a large-lobed, coralloid structure.

Caecomyces was erected to contain species with uniflagellate zoospores that have a vegetative stage consisting of a single sporangium borne on a coralline rhizoidal system carrying bulbous structures and sometimes fibrillar roots. The type species, *equi*, from the feces of the horse (36), is the only species fully described.

Colonies of *Caecomyces equi* growing on cellobiose in vitro are granular in appearance, very similar to those of *Sphaeromonas communis* (Fig. 6d), but are relatively slow-growing compared to *Neocallimastix* spp. and *Piromyces* spp.

Sphaeromonas Liebetanz (= *Sphaeromonas* sensu strictu Orpin [16]).

Zoospore spherical to ellipsoidal to amoeboid, motile with a single flagellum. Nucleus central; organelles distributed throughout the cytoplasm, hydrogenosomes near the kinetosomes. Ribosome aggregates present. Plasmalemma coated with elaborately organized fibrous material. Vegetative growth monocentric, holocarpic with endogenous development, consisting of a poorly branched thallus bearing the sporangium and one or more spherical structures (16,37).

The generic name *Sphaeromonas* was first used by Liebetanz (4) in his descriptions of flagellate protozoa from the rumen. Orpin (16) used the name *Sphaeromonas communis* for an isolate with zoospores similar to those described by Liebetanz as the flagellate *S. communis*. The description by Liebetanz has since proved illegitimate, since we now know these organisms to be fungi, and although it was proposed that the new genus *Caecomyces* should include *Sphaeromonas* (36), the current author regards *Sphaeromonas* as distinct from *Caecomyces*. *Sphaeromonas* stains isolated from the rumen (Fig. 8) are not known to produce fine-filamentous as well as coralline or bulbous rhizoidal structures, and *Caecomyces* is not known to produce more than a single rhizoidal bulbous structure per plant, as does occur in *S. communis* (Figs. 8a, 8c–e). Close examination at the level of molecular systematics is required to determine the relatedness of the two genera.

The sporangia may be attached by a short sporangiophore to the spherical rhizoidal structures (Figs. 8a,c,d) or by an elongated sporangiophore (Fig.

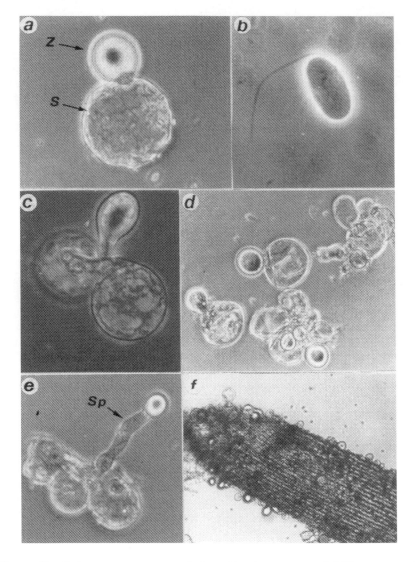

Figure 8 *Sphaeromonas communis* from the sheep rumen. (a) The monocentric thallus is shown, which bears a single, spherical rhizoidal structure (S). (b) The zoospore normally bears in single flagellum. (c,d,e) Thalli with multiple spherical structures are common in cultures in vitro. The sporangia may be borne on a short sporangiophore (c) or an elongated sporangiophore (e). (f) Plant tissue invaded in vitro is heavily colonized, with the sporangia projecting from the external surface and the rhizoidal system mostly within the plant tissues. Phase-contrast microscopy. (a) ×375; (b) ×1250; (c) ×375; (d) ×180; (e) ×375; (f) ×75.

8e). The organism colonizing plant tissue is shown in Fig. 8d; the uniflagellate zoospore is shown in Fig. 8b.

A detailed study of *Caecomyces (Sphaeromonas) communis* showed that at certain times the rhizoids of this strain lengthen to form filaments that may contain nuclei (49).

Colonies of *Sphaeromonas communis* grown in vitro (Fig. 6d) are slow-growing compared with *Neocallimastix* spp. and *Piromyces* spp. and are granular in appearance. Colonies do, on occasion, produce a few thick, blunt, radially developing rhizoids considerably shorter than the radial rhizoids of colonies of *Neocallimastix* and *Piromyces*.

Piromyces Gold (36) ex Li, Heath and Bauchop (38) emended.

Anaerobic gut fungi. Sporangium ovoid, elongate, irregular, papillate or nonpapillate; borne on a single, extensively branched rhizoidal system, with or without subsporangial swellings. Zoospore spherical to ovoid, posteriorly uniflagellate, rarely biflagellate to quadriflagellate. Type species, *Piromyces communis*.

This genus was erected to embrace (monocentric) species with principally uniflagellate zoospores (Fig. 9c), often pyriform in shape, and a vegetative stage with a single sporangium borne on a single, branching rhizoidal system. It includes the strains previously known as *Piromonas communis* (14,21; Fig. 9a). Liebetanz (4) published a description of a singly flagellate organism corresponding to the fungal zoospores found in culture by Orpin (14), who subsequently adopted the name *Piromonas communis* for his isolate. Since the flagellates were fungal and not protozoal in organ, the description by Liebetanz was illegitimate, and the new generic name of *Piromyces* was proposed (36), but the description was subsequently emended (33). *P. communis* (14) from the sheep rumen is the type species, with two other species, *mae* and *dumbonica*, described from the feces of the horse and the Indian elephant, respectively (38).

The zoospores of *Piromyes* are principally uniflagellate (Fig. 9c), though individual zoospores may carry up to four flagella (14,33,43). In exceptional circumstances in old cultures in the laboratory, the same species may produce deformed zoospores that have not differentiated properly within the sporangium contents, with the subsequent release of a zoospore mass possessing many flagella. Barr and his group (33) observed both endogenous and exogenous development in strains of *Piromyces*. Similar exogenous growth in aging cultures of *Piromyces* has been observed by the author; exogenous growth is rare or absent in young cultures growing in fresh media. Because of this, exogenous development may not represent normal development in *Piromyces*. This facet of *Piromyces* development requires further detailed examination before it can be claimed to be a characteristic of the genus.

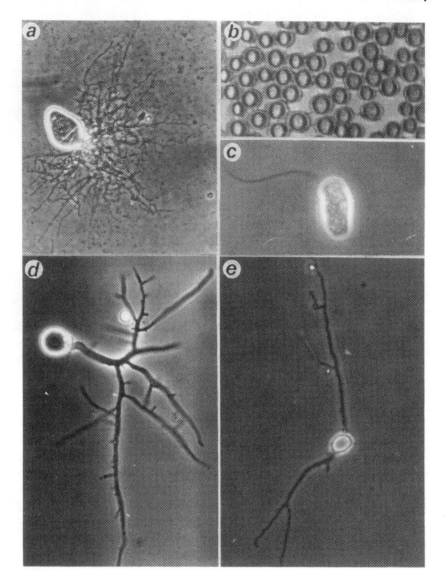

Figure 9 Aspects of *Piromyces* biology. The vegetative stage is monocentric, but the rhizoids may be much divided as in *P. communis* **(a)** or similar to *Neocallimastix* spp. as in strain NF from sheep **(d)**. The zoospore **(c)** may develop by the production of a single rhizoid **(d)** or two rhizoids **(e)**. An isolate of *Piromyces* from the horse cecum developed dormant spherical structures **(b)** in laboratory cultures. **(c)**, **(d)**, and **(e)** are of *Piromyces* strain NF from the sheep rumen. **(a)** ×200; **(b)** ×220; **(c)** ×1250; **(d)** ×500; **(e)** ×500.

At least two species of *Piromyces* occur in the rumen of both sheep and cattle. They are *P. communis*, the type species, and another larger species designated NF and F (22,23; Figs. 9c–e, 10b), which has yet to be legitimately described. The latter has a vegetative stage similar to *Neocallimastix* but has uniflagellate zoospores. Unlike *Neocallimastix frontalis* and *patriciarum*, the sporangiophore is much constricted near the base of the sporangium.

Other species that are probably species of *Piromyces* have been isolated from the cecum or feces of the horse, the Indian and black rhinoceros, the Indian and African elephant, and the mara, from the forestomach of the eastern grey kangaroo, and from feces of the wallaroo (Tables 1, 2).

The morphology of colonies of *Piromyces communis* growing in vitro on cellobiose is similar to that of *Neocallimastix* spp. in that the central core of sporangia is surrounded by radially growing rhizoids, but the colonies are smaller than those of *Neocallimastix* spp. A culture of *Piromyces* isolated from the horse cecum produced spherical, highly refractile structures (Fig. 9b) near the meniscus in liquid culture (25). The nature of these structures is unknown, but they may represent survival stages.

The vegetative stage of another horse cecum isolate of *Piromyces* is shown in Fig. 10a, exhibiting a less branched rhizoidal system than *P.communis* (Fig. 9a). This is constant difference and is not caused by the presence of methanogenic bacteria (Fig. 9a) in the coculture depicted.

Table 1 Key for Identifying Anaerobic Fungi to the Generic Level

1	Vegetative growth/monocentric	2
	polycentric	5
2	Zoospores with more than 7 flagella; usually 7 to 15	*Neocallimastix*
	Zoospores with 1 to 4 flagella	3
3	Vegetative growth with bulbous rhizoids	4
	Vegetative growth without bulbous rhizoids; rhizoids filamentous	*Piromyces*
4	Only bulbous rhizoids and (in old cultures) thick filaments present.	*Sphaeromonas*
	Some fibrillar rhizoids present, or rhizoid coralline	*Caecomyces*
5	Zoospores uniflagellate	*Ruminomyces, Anaeromyces*
	Zoospores multiflagellate	*Orpinomyces*

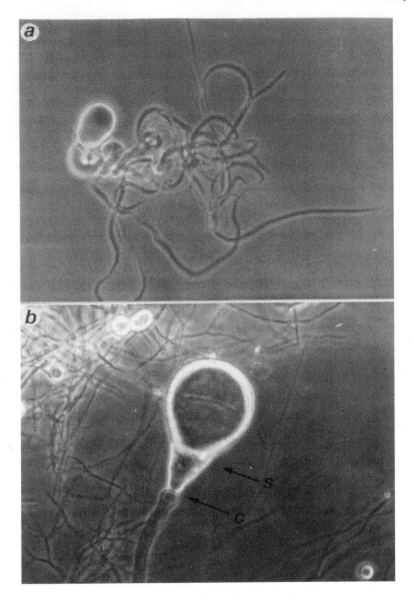

Figure 10 Strains of *Piromyces* spp. from the horse cecum (**a**) and strain NF from sheep (**b**). Both are more similar in morphology to *Neocallimastix* than *Piromyces communis* (Fig. 9a). Strain NF from sheep consistently possesses a constriction (**c**) between the sporangiophore and the sporangium, and a septum (S) separating the sporangium contents that develop into zoospores from other material in the base of the sporangium. (**a**) ×370; (**b**) ×370.

Table 2   Occurrence of Anaerobic fungi in Ruminants

Species of animal	Types of fungi found	Ref.
Domestic sheep (*Ovis aries*)	N,P,S,O	9,14,16,76
Domestic cattle (*Bos taurus*)	N,P,S,O,A	12,32,33,50
Domestic cattle (*Bos indicus*)	N,P	52,72
Domestic goat (*Capra hircus*)	N,S,P	27
Barbary sheep (*Ammatragus lervia*)	S,N	27
Gaur (*Bos gaurus*) (feces)	S,N	62
Musk Ox (*Ovibos moschatus*)	S,P	27
Moufflon (*Ovis ammon musimon*)	S,N,P	27
Water Buffalo (*Bubalus arnee*)	S,N,O	52,122
Red Deer (*Cervus elephus)*	Sp	15
Impala (*Aepyceros melampus*)	Sp	62
Norwegian reindeer (*Rangifer tarandus*)	N	27
Svalbard reindeer		
(*Rangifer tarandus platyrhynchus*)	N	64

N = *Neocallimastix* spp.; P = *Piromyces* spp. S = *Sphaeromonas* spp.; O = *Orpinomyces* spp.; A = *Anaeromyces* spp.; Sp = unidentified sporangia seen in rumen contents.

C. Polycentric Genera and Species

Orpinomyces Barr and Kudo (33).

Polycentric rhizomycelium with terminal or intercalary branched sporangiophore complexes. Zoospores multiflagellate. Obligately anaerobic fungi in the digestive tract of herbivores. Type species *Orpinomyces bovis.*

This genus was erected to accommodate polycentric species with multi-flagellate zoospores and exogenous development, with the type species *bovis* having been isolated from cattle (33). A second species that should properly be placed in this genus, and is placed so here, is *joyonii* (Fig. 11), originally included in the genus *Neocallimastix* (48).

On germination, the contents of the zoospore move into the germ tube, in contrast to the monocentric species (in which genera the zoospore enlarges to form the sporangium); the germ tube then develops into an extensive polynucleate rhizomycelium (48) on which the sporangiophores, and subsequently sporangia, develop.

Colonies of *Orpinomyes* spp. growing on cellobiose in culture in vitro may develop radially up to 1 cm in diameter, both in liquid an in agar-containing media (Fig. 6e), and are considerably larger than those of the monocentric species.

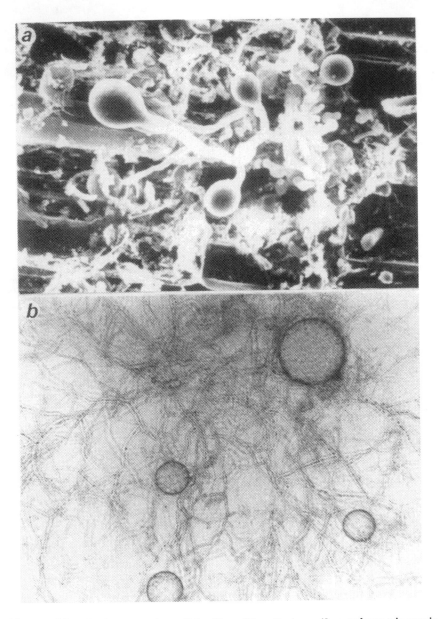

Figure 11 *Orpinomyces joyonii* (= *Neocallimastix joyonii*), a polycentric species from sheep, showing multiple zoospores on a single, much-branched rhizoidal system. Growth on plant tissues **(a)**, demonstrated by scanning electron microscopy, and in medium containing soluble substrates **(b)**, photographed by light microscopy, is shown. **(a)** ×280; **(b)** ×250. (Courtesy of G. Fonty.)

Ruminomyces Ho (50).

Rhizomycelium complex, polycentric. Hyphae much-branched, large, often with constrictions, or fine and rootlike. Sporangiophore erect, solitary, unbranched, arising laterally or terminally from the hyphae. Sporangia solitary, broadly ellipsoidal to fusiform, usually with an acuminate apex. Zoospore uniflagellate, spherical to variable in shape.

The genus *Ruminomyces* was erected (50) to embrace uniflagellate polycentric species originally found in cattle (*Bos indicus* and *Bos taurus*) and water buffalo (*Bubalus bubalis*) (51,52). The type species for the genus is *Ruminomyces elegans*, from cattle (50).

Anaeromyces Breton et al. (53).

Extensive polycentric thallus, with a highly branched, cenocytic, polynucleate rhizomycelium. Elliptical zoosporangia, 29 (54) 120 × 15 (30) 75 μm, with an apical mucro. Sporangiophores 31 (58) 83 × 4 (9) 16 μm. Spherical zoospores, 7.5–8.5 μm in diameter, with a single flagellum up to 30 μm. Type species *Anaeromyces mucronatus*.

The genus *Anaeromyces* was, like *Ruminomyces*, erected to contain polycentric species with uniflagellate zoospores. The type species of both genera were isolated from cattle, and both species have mucronate sporangia. There do seem to be some morphological differences between the two isolates, such as that *Ruminomyces* rhizomycelia are commonly constricted, and more work is necessary to determine the precise relationships of these two genera.

D. Identification to the Genus Level

A simple key is presented (Table 1) that may be used to position a new isolate of an anaerobic fungus, isolated from either a foregut fermenter or a hindgut fermenter, among the known genera.

III. EXAMINATION OF SAMPLES

A . Fresh Gut Contents

Samples of gut contents can be examined fresh, while still warm, using an inverted microscope or a normal binocular microscope preferably fitted with phase-contrast optics, at a magnification of × 100. It is preferable to use a heated (39°C) microscope stage that will retain the viability and movement of the zoospores. If anaerobic fungi are present it is usually possible to see either free-swimming zoospores (Fig. 1) or pieces of plant tissue infested with the vegetative growth of the fungi (Fig. 2). Sporangia are often seen more easily at cut ends of plant fragments. It is not difficult, particularly with a little experience, to differentiate between anaerobic fungal zoospores and other organisms. The zoospores are considerably smaller (5–25 μm long) than the majority of ciliate protozoa (see Fig. 1c), and although they are similar in size to true flagellate protozoa (e.g., *Pentatrichomonas* spp. [10]), they can be readily distinguished by their morphology, flagellation, highly refractile appearance, and characteristic swimming motion. The swimming of the fungal zoospores is in a highly directed wriggling motion reminiscent of a tadpole, gyration, or a rolling motion. Zoospores of *Neocallimastix* spp. usually either gyrate rapidly or swim rapidly for quite large distances in straight or curved lines if observed at 39°C, but may also suddenly stop and gyrate, or change course.

Zoospores of rumen anaerobic fungi can be differentiated from the rumen large bacteria *Magnoovum eadii* (54) (Fig. 1a) and Quin's Oval (55) (Fig. 1b) by the possession by the zoospores of one or more flagella that can be observed in the light microscope, their morphology, and their more vibrating swimming motion. In contrast to the more flexible cell wall of the zoospores, which allows them to take on an ameboid form, the large bacteria have a rigid cell wall. Although the flagellate *Magnoovum eadii* is similar in size to the body of some of the zoospores of the anaerobic fungi and swims rapidly, no flagella are visible in the light microscope. The swimming motion of *Magnoovum eadii* is smooth, with slow rotation about the long axis. It does not show the somewhat jerky wriggling movement shown by fungal zoospores. Quin's Oval is smaller than most anaerobic fungal zoospores, swims relatively slowly, and again, no flagella can be seen under the light microscope. Large strains of *Selenomonas ruminantium* (56) (Fig. 1b) are normally constantly curved, and although the lateral tuft of flagella may occasionally be seen in the light microscope at high magnification, its movement in rumen contents is normally weak, nondirected, and spinning. In contrast to the fungal zoospores, which stain yellow with iodine, the large bacteria stain medium to deep brown (54,55,56), except when starved.

The flagellar bundle of the zoospores of *Neocallimastix* spp. can be observed at a magnification of ×100, and when the zoospore is swimming the flagella beat as a single locomotory organelle that trails posteriorly. However, the side of the cell where the flagella originate can often be seen to retain a lateral rather than a posterior position. The individual flagella of which the bundle is composed, and the single flagellum in uniflagellate species, are difficult to see in living preparations except at a magnification in excess of ×400. If a living preparation is allowed to cool to ambient temperature, the motion of the zoospores becomes less and the flagellar bundle of the *Neocallimastix* spp. zoospore relaxes, and it becomes possible to differentiate the individual flagella within the bundle, particularly near the cell surface. If the preparation cools to ambient temperature for more than a few minutes, the zoospores may stop moving or die and will be difficult to detect microscopically.

Occasionally during observations of fresh rumen contents, zoospores are observed that cease swimming and commence movement as ameboid cells (Figs. 4d, 9c), crawling over the surface of the glass slide or digesta fragments in the preparation (9). This behavior is more commonly observed in cultures in vitro. Zoospores of the uniflagellate species usually swim more slowly, with the flagellum posteriorly oriented, than those of *Neocallimastix* spp., but not always with the same directness. When observed under phase-contrast illumination in the light microscope, they are more refractile than the flagellate protozoa and swim more actively and more directionally.

Fresh preparations of gut contents can be stained with iodine (Fig. 2c) (57) or lactophenol cotton blue (58) to enable the vegetative growth of anaerobic fungi to be visualized more easily on digesta fragments. The fungal sporangia and rhizoids, where thick enough, stain deep brown with iodine and blue with lactophenol cotton blue. The zoospores stain yellow with iodine and pale blue with lactophenol cotton blue, but the intensity of staining of the zoospores with iodine is rarely sufficient for them to be picked out among the fragments of digesta (59).

When searching preparations for vegetative growth care is necessary to differentiate between plant tissues that have been infested with aerobic fungi before the animal eats them and anaerobic fungi from the gut. However, the presence of sporangia attached to rhizoids that penetrate the plant tissue usually means that anaerobic fungi are present. These are identifiable by their highly refractile appearance or by staining with iodine or lactophenol cotton blue. Aerobic fungal contamination of temperate grasses is often topical, and except when it is heavy, fungal sporangia are infrequently observed.

One of the best ways to observe living zoospores in rumen contents is to starve the animal for at least 8 h, then to feed it and remove a sample of rumen contents for microscopic examination 15–60 min later. Under these conditions, the population density of zoospores found after feeding may be as high as 1500 times the prefeeding population density (9,13) and offers the best chance to observe the free-swimming zoospores. In wild animals or animals at pasture such control is not possible, and free-living zoospores may not be observed despite the presence of anaerobic fungi. This, however, is rare, and if anaerobic fungi are present, the zoospores are usually evident in samples of rumen contents taken at any time of day.

Sporangia can usually be observed on the cut ends, at the stomata, and at damaged regions of the plant surface of infected plant fragments (Fig. 2). It is usual for the anaerobic fungi to be found principally on, or within, the larger plant particles. It is nor normal for the plant particles to be uniformly infected by the fungi, and this is not surprising, since it is the freshly ingested plant fragments that are invaded by the fungi. The smaller plant fragments have been in the rumen for longer periods and may previously have hosted populations of fungi, and the remnants of rhizoids similar to those of anaerobic fungi may sometimes be seen on these fragments. In addition, different tissues are invaded to differing extents (13). A region that is often overlooked but is often particularly rich in anaerobic fungal growth is the interior of hollow grass or straw stem fragments.

Visual examination alone is not always adequate to determine the presence or absence of anaerobic fungi (59), and it may be necessary to resort to a cultural method (Chapter 3). The method of choice, since it is the most sensitive currently available, is the most-probable-number (MPN) method described by Theodorou and his co-workers (24), which is sensitive enough for the detection of as few as 10–100 thallus-forming units per gram of digesta; it may be impossible to detect fungi in this range of population densities by visual methods because of the particulate nature of the digesta. However, for obtaining isolates representative of the whole population, dilution and subsequent culture in an agar medium (17,28,60) is preferred.

B. Fixed Gut Contents

Contents of the intestinal tract may be fixed with formaldehyde, glutaraldehyde, or iodine (57,61) while still warm, and examined later. In samples fixed with iodine, a common method of fixation for subsequent determination of protozoal species or population densities, the sporangia of the anaerobic fungi stain deep brown, with the rhizoids staining lightly; the zoospores stain yellow. Fungi in material fixed in formaldehyde or glutaraldehyde will also stain with iodine or lactophenol cotton blue (58), to reveal fungal growth. Again, examination of control preparations of stained dietary material is necessary to determine fungal contamination in the diet and hence rule out false positives. The same staining methods can be applied to fresh preparations.

IV. DISTRIBUTION IN NATURE

A range of uniflagellate and multiflagellated anaerobic fungi have been isolated from or observed to be present in large number of ruminants (Table 2) and nonruminant herbivores (Table 3). To date, however, only uniflagellate monocentric species have been found in the alimentary tracts of nonruminant herbivores; multiflagellate species have been isolated only from ruminants and from camels (27). Polycentric species have been isolated from cattle and water buffalo (33,48,50,51,52) and sheep (C. G. Orpin, unpublished results). Attempts have been made (25) to isolate the multiflagellate *Neocallimastix equi* observed in horse cecum contents (6), but no cells resembling this species were seen in the cecum contents of a small number of horses

Table 3 Occurrence of Anaerobic Fungi in Nonruminant Herbivores

	Fungi	Site	Ref.
Camelidae			
Dromedary (*Camelus dromedarius*)	N	Fs	63
Llama (*Lama guanicoe*)	N,P	Fs	63
Pessidactylia			
Horse (*Equus caballus*)	P,C	Ce,Fe	15,36,38
Donkey (*Equus asinus*)	P,C	Fe	63
Zebra (*Equus caballus*)	Sp	Fe	63
Indian elephant (*Elephas maximus*)	P	Fe	38,65
African elephant (*Loxodonta africana*)	P,C	Fe	15,65
Black rhinoceros (*Diceros bicornis*)	P	Fe	65
Indian rhinoceros (*Rhinoceros unicornis*)	P	Fe	65
Rodentia			
Mara (*Diplochotis patagonum*)	P	Fe	65
Guinea pig (*Cavia aperea*)	S	Ce	16
Macropodidae			
Eastern grey kangaroo (*Macropus giganteus*)	P	Fs	63
Wallaroo (*Macropus robustus*)	P	Fe	63
Redneck wallaby (*Macropus rufogriseus*)	Sp	Fs	63
Swamp wallaby (*Macropus bicolor*)	Sp	Fs	63

N = *Neocallimastix* spp.; P = *Piromyces* or *Piromyces*-like species; C = *Caecomyces* spp.; S = *Sphaeromonas*-like strains present; Sp = sporangia and vegetative growth observed in faeces, or in organ contents; Fs = forestomach; Ce = cecum; Fe = feces.

examined in England (25) or in three animals from Australia (C. G. Orpin, unpublished), so no conclusions can yet be made about the status of this organism, despite the statement that "*C. equi* du colon des Equides (Hsuing, 1929) est sans doute synonyme de *C. frontalis*" (8).

Bauchop (28,63) reports evidence for anaerobic fungi in the forestomach of several macropod marsupials. The existence of anaerobic fungi in the forestomach of the Eastern grey kangaroo (*Macropus giganteus*) has been confirmed (63), and they have been isolated from the feces of the wallaroo (*Macropus robustus*) but not the feces of the Eastern grey kangaroo (C. G. Orpin and M. K. Theodorou, unpublished). All strains that were isolated from both animals were monocentric, uniflagellate strains similar to *Piromyces*. It is also interesting to note that uniflagellate species have also been isolated from two rodents: the mara and, in a single instance, the guinea pig (16); examination of other (large) rodents may yield other hosts of anaerobic fungi.

Animals in which no evidence for gut anaerobic fungi has been found (27; C. G. Orpin, M. Teunissen, unpublished) include the roe deer (*Capreolus*), muntjac deer (*Muntiacus muntjac*), red brocket deer (*Mazama americana*), hippopotamus (*Hippopotamus amphibius*), pigmy hippopotamus (*Choeropsis liberiensis*), giant panda (*Ailuropoda melanoleuca*), domestic pig (*Sus scrofa*), rabbit (*Oryctolagus cuniculatus*), european hare (*Lagus europaeus*), golden hamster (*Mesocricetus aureus*), Mongolian gerbil (*Meriones unguiculatus*), mouse (*Mus domesticus*), rat (*Rattus norvegicus*), coypu (*Myocaster coypus*), koala (*Phascolarctos cinereus*), common brush-tail possum (*Trichosurus vulpecula*), minke whale (*Balaenoptera acutorostrata*) (S. D. Mathiesen, personal communication), and marine iguana (*Amblyrhynchus cristatus*). Rumen, forestomach, or cecal samples, as appropriate, were examined in all species except the hippototami and giant panda, for which only fecal material was available. Cultural methods were used to detect anaerobic fungi except in the instance of the red brocket deer and marine iguana, for which only fixed material was available. It is interesting to note that the ruminants apparently lacking anaerobic rumen fungi are all small, with a consequent high ruminal flow rate, and are selective feeders eating young browse or (red brocket deer) palm nuts. With some species the number of samples examined was small, so anaerobic fungi may yet be found in some of these animals.

V. LIFE CYCLES OF ANAEROBIC FUNGI

Warner (66) observed that the population density of *Neocallimastix* flagellates (polymastigotes) in animals fed once daily showed a large increase shortly after the host animal was fed in comparison with the prefeeding population density. It was this observation that prompted Orpin (59) to seek an explanation and eventually resulted in the discovery of the rumen anaerobic fungi. Most rumen organisms show a decrease in population density after the animal eats, owing to a dilution of rumen contents with dietary water and saliva. The phenomenon could be repeated in vivo by adding an acetone extract of dietary plant material to the rumen (67) in place of the diet, and subsequently by the addition of hemin (68). It was deduced that hemes present in the plant material triggered zoospore differentiation and release into the rumen liquor. A source of heme is essential for growth of *Neocallimastix patriciarum*, since the organism is unable to insert iron into the tetrapyrrole ring to synthesize heme. Growth and zoosporogenesis is supported by a range of hemes, including both reduced and oxidized forms. In vivo, at the low redox potential existing in the rumen, oxidized forms of heme will be reduced (68). It is therefore likely that reduced forms of heme, rather than oxidized forms, induce zoosporogenesis and zoospore release. Thus all plants, living or dead, that contain heme-containing compounds would trigger the differentiation and release of zoospores. This ensures that zoospore production occurs at the time fresh food enters the rumen. This characteristic has been employed with some success to synchronize zoospore production in liquid, heme-limited cultures, but the yield of vegetative growth and production of zoospores was small.

The response to dietary heme is not the only factor involved in zoospore production and release, since zoospores at low population densities may be observed in rumen contents at any time of day in animals fed once a day. In addition, when stem tissues of straw are cut open, it is common to find fungi of all stages of maturity inside. In rumen contents of free-ranging animals, fungal sporangia of all stages of maturity can also be observed.

In animals fed once daily there is a surge in zoospore production 15 min to 1 h after feeding as compared with the prefeeding level (9,14,16). In free-ranging animals and those fed more frequently, the situation is not clear, although zoospores may be seen in rumen contents at any time of day. In animals fed once a day, zoosporogenesis can be induced on feeding, then after about another 8 h (67,68), suggesting that the fungi need to be at least 8 h old before they are mature enough to produce zoospores. However, this age most likely varies with the diet of the animal and the rate of growth of the fungi. Small sporangia of *Neocallimastix* spp. from the sheep rumen have been observed to release as few as two zoospores (67), and large sporangia as many as 88 zoospores (69) or even more (Fig. 7), indicating that the mass of the sporangium is unimportant in relation to maturity of the plant. However, the production of only two zoospores by small sporangia may be a result of their having been broken from the rhizoidal system, triggering their maturation.

If rumen contents are strained roughly and forcefully through cheesecloth, a number of fungal sporangia are broken from the rhizoids (Fig. 2b). They can be differentiated from protozoa by their rounded shape and usually by the presence of an attached length of rhizoid. If this suspension is treated with an acetone extract of oats after evaporative removal of the solvent and observed under phase-contrast microscopy on a heated microscope stage, differentiation and release of the zoospores can be observed. Release of the zoospores occurs most frequently through regions of the sporangium wall that have dissolved (9,69) (Figs. 4e, 7), although some authors suggest that in some species papillae exist through which the zoospores are released. Papillate sporangia appear to have been reported only from some *Piromyces* spp. (38).

The zoospores of *Neocallimastix patriciarum* are attracted to plant tissues that have been freshly ingested by the host animal, by chemotaxis to soluble carbohydrates diffusing from those tissues (70). The common soluble carbohydrates of plant tissues, glucose, fructose, and sucrose, all elicit chemotaxis of the zoospores and act synergistically at low concentrations. Such synergy may result in tissue selectivity of the zoospores, with those of *N. patriciarum* being more attracted to inflorescence tissues, richer in soluble carbohydrates than other tissues. In practice this is important only if the animal is consuming forage that is flowering, which is rare in managed pastures. The threshold concentration that elicited chemotaxis in buffer solution was about one tenth of that eliciting chemotaxis in rumen fluid, which was in the range of 10^{-4}–10^{-5} M.

Attraction of zoospores to fresh plant tissues can be observed if the tissues are added to freshly collected rumen fluid, containing zoospores of anaerobic fungi, and examined with a light microscope using a heated (39°C) stage. The zoospores can be seen to swim toward the plant tissue, together with some protozoa and large bacteria, and to locate the cut ends of the tissue or the stomata. Here they cease swimming and assume the ameboid phase and, after crawling on the tissue, locate a suitable site for encystment. Attachment is mediated by the production of fibrillar material (Fig. 5a;) (35). The zoospore ceases movement and assumes a rounded shape, and the cell wall becomes thickened to form a cyst (Fig. 5f). After encystment, normally a single rhizoid is produced, which penetrates the plant tissue through the damaged area or a stoma.

Growth and extension of the rhizoid through the plant fragment usually results in attachment of the rhizoid to the more lignified vascular tissue within the plant fragment, although the entire rhizoidal system may exist in the less lignified tissue, apparently not attached to the vascular bundles. This occurs occasionally with the smaller rhizoidal species such as *Piromyces* spp. *Sphaeromonas communis* and *Caecomyces equi*, on the other hand, may enter the tissue and grow to produce tumescences, rupturing the surface of the tissue and filling the cell contents with fungal tissue. It is likely that the "spherical bodies" of Orpin (16) actually fill vacuoles within the plant cells in vivo (71). In mature plants, particularly of *Neocallimastix* spp., with extensive rhizoidal systems, the rhizoid can be stained with Lugol's iodine, tracked visually with a light microscope, and observed to penetrate the vascular tissue and to travel longitudinally along the xylem cells. Appressoria are sometimes observed where the rhizoid is attached to the plant cell walls (72). Fungi can frequently be observed in rumen contents attached by a length of rhizoid to an isolated vascular bundle; this is probably due to dissolution of the surrounding, more readily digested tissues by ruminal digestion, before maturation of the fungus and release of the zoospores. Alternatively, particularly if digestion of the tissue is not extensive, the sporangium may have developed exogenously from a rhizoid tip.

It has not yet been determined whether the rhizoids of the anaerobic fungi enzymatically hydrolyze the plant cell wall before penetration. Although there is evidence that dissolution of the cell way may occur, other evidence suggests that the rhizoids may penetrate through existing pores in the cell wall. Frequently appressoria can be observed at the points of contact between the rhizoids and plant cell walls where the rhizoid penetrates the wall (72).

The growth of the rhizoidal system has been studied in *Neocallimastix hurleyensis* (69). In this species, the rhizoid lengthens extensively before the zoospore cyst enlarges. Considerable branching and enlargement of the rhizoid diameter then occurs concomitantly with expansion of the sporangium. This is similar to that found in *Neocallimastix patriciarum* (44) (Fig. 4a). The life cycle of monocentric rumen anaerobic fungi cultured in vivo is in the range of 24–32 h (9,15,69).

During reproduction numerous nuclear divisions are required. Studies of the somatic structures of anaerobic fungi have revealed that in monocentric species the nucleus of the young plant remains within the developing sporangium, with subsequent nuclear division during the maturation of the sporangia (73,74). Nuclei of the polycentric *Orpinomyces joyonii* (= *Neocallimastix joyonii*) occur within the rhizoidal system and increase in number before the development of sporangia.

Since viable anaerobic fungi have been found in feces of many ruminants and hindgut-fermenting herbivores (24,25,63,75,76), it is likely that transfer between animals, including transfer from mother to offspring, may be mediated by fecal contamination. Anaerobic fungi have also been isolated from the esophagus of ruminants (62,76), and transfer may be effected by licking, by eating from contaminated food or food bins, or by aerosol. Establishment of populations of rumen fungi in young ruminants is capricious in the preruminant animal (77). Populations can be established in young, isolated animals by direct inoculation with laboratory-grown cultures (77). Young animals will also become infected with rumen fungi with no direct contact with other animals, presumably by aerosol or dust contamination from nearby animals (78).

In hindgut-fermenting herbivores that harbor populations of anaerobic fungi, infection of young animals is presumably by coprophagy or the ingestion of feces-contaminated materials. In order for the ingested fungi to inoculate the cecum or large intestine they must successfully make the transit of the abomasum and small intestine. The thin-walled zoospores are unlikely to survive this passage, and some vegetative growth may be digested in the alimentary tract. However, the thick-walled nature of the sporangia may allow a successful transit, or alternatively, a resistant structure may develop. Rumen fungi are known to survive transit of the alimentary tract of ruminants and have been detected in all sections of the alimentary tract (24). The development of an oxygen-resistant survival stage in dried feces has been demonstrated, but its nature is still unknown. A resistant sporangium or cyst was postulated, and melanized "resistant sporangia" have now been demonstrated in cultures (121). One isolate from the cecum of the horse, morphologically similar to the rumen *Piromyces* spp., produced shperical, thick-walled structures smaller than sporangia but larger than the zoospores, believed to be a resting stage, abundantly on the wall of culture tubes in aging cultures (25). The thick-walled "warty spores" observed in water buffalo rumen contents (122) may also represent resting or survival stages. In contrast to this, although anaerobic fungi occur in the forestomach of the eastern grey kangaroo, they have not been isolated from the lower alimentary tract and may not generate resistant stages (C. G. Orpin and M. K. Theodorou, unpublished).

The food of hindgut-fermenting herbivores is subject to digestion in the upper alimentary tract before entry into the hindgut. Most of the readily metabolized and soluble components will have been removed. Thus the massive induction of zoosporogenesis by the release of heme, seen in the rumen, may not occur in the hindgut.

VI. CHEMISTRY

A. Carbohydrates

Analysis of vegetative growth of anaerobic rumen fungi indicates that up to 40% of the dry weight of the cell wall consists of chitin (11), and about 11.8% of the dry weight of *Neocallimastix patriciarum* and 7.8% of that of *Piromyces communis*. Other strains of *Piromyces* (79) contain higher levels of chitin, suggesting some diversity among ruminal *Piromyces* isolates. Chitin is only partly digested in the rumen (80), and thus some of the residual cell wall remaining in the rumen after the release of zoospores may remain undigested. The survival of the cell walls of the vegetative stage of rumen fungi has not been studied, although it is known that the digestibility of rumen fungi in rumen fluid is high (81). No other rumen organisms are known to contain chitin, and the assay of chitin as a marker has been employed to measure fungi associated with digesta in vitro, in fermentation experiments with preparations inoculated with rumen fluid (82), and to measure the fungal biomass in the rumen (83). The latter met with little success because of interference with the reaction by components in the reaction mixture and the long incubation times necessary for chitin hydrolysis. However, a specific probe for chitin synthase (84), presumably absent from other rumen organisms, may prove of value in estimating fungal biomass. Other specific probes are likely to be developed from species- or group-specific nucleotide sequences.

Glycogen-like storage carbohydrate is also present in the zoospores and vegetative growth of *N. patriciarum* (35,37), *N. frontalis* (32), *Piromyces communis*, and *Spaeromonas communis* (37). In *N. frontalis* it may represent 35–40% of the dry weight (79). Since all anaerobic fungi stain brown with iodine, all probably contain a storage polysaccharide similar to glycogen.

B. Sugar Alcohols

Acyclic polyols, synthesized by the majority of the Eumycota, are taxonomically significant (85,86) and may be used to confirm taxonomic relationships. The anaerobic fungi *Neocallimastix patriciarum* and *Piromyces communis* contain glycerol as the only acyclic alcohol (87). This is in contrast with the other chytridiomycete fungi *Allomyces arbuscula* and *Blastocladia emersonii*, which contain principally mannitol and arabitol as the major acyclic alcohols (88). Since the anaerobic fungi occupy quite a different habitat from these species, this chemotaxonomic difference supports the placing of the anaerobic species in a different taxon.

C. Lipids

The lipids of *Piromyces communis*, *Neocallimastix patriciarum*, and *Neocallimastix frontalis* have been examined (89,90). The lipid composition of all three species was similar in many respects. The major phospholipids were phosphatidylethanolamine, phosphotidylcholine, and phosphatidylinositol. Sphingolipids, glycolipids, plasmalogens, and phosphoryl lipids were apparently absent. Synthesis of long-chain lipids was achieved from glucose and acetate (89), and short- and long-chain fatty acids could be taken up from the growth medium and incorporated into complex lipids.

The fatty acid profiles reflected the anaerobic growth conditions required by the fungi. No polyenoic acids, which require oxygen for their synthesis and are common in aerobic fungi, were detected. High levels of monoenoic acids with chain lengths of up to C24 were detected in all three species. The most abundant was oleic acid (18:1 *cis*), which represented 70% of the total of the *n*-unsaturated fatty acids in *N. patriciarum*. Evidence was obtained that oleate was synthesized by chain elongation of saturated acids to stearate, which was then desaturated to oleate. This reaction implies an oxygen-independent desaturase, since rigorous exclusion of oxygen from the system still resulted in stearate desaturation. In addition, no evidence for the presence of cytochromes, necessary if oxygen was the terminal acceptor, were found.

The neutral lipid fraction of *P. communis* and *N. patriciarum* contains squalene and a triterpenoid, tetrahymenol (89). Sterols were not detected, and it is assumed that the place of sterols in cell membranes of these species was taken by squalene and tetrahymenol. In the absence of oxygen, sterol synthesis would be unlikely, since oxygen is essential for squalene cyclization in aerobic organisms (91). Nystatin and amphotericin B, antibiotics known to inhibit sterol synthesis, were ineffective in inhibiting the growth of *Neocallimastix hurleyensis* (69), confirming that sterol synthesis in these fungi probably does not occur.

D. Proteins and Amino Acid Composition

The protein content of *Neocallimastix* spp. and *Piromonas communis* is high, in the range 25–30% of the dry weight (81,92), with amino acid profiles similar to those of casein and lucerne fraction 1 protein (81), suggesting that these fungi may contribute substantially to the supply of amino acids to the animal. It is not known what proportion of the fungal biomass that leaves the rumen is available to the animal, but zoospores, at least, would be hydrolyzed in the abomasum. Fresh rumen fungi infused into the abomasum were also extensively digested (92), with an average fractional digestibility of the constituent amino acids of about 0.9, higher than the value of 0.74 given for total rumen microbial protein (93), indicating that the manipulation of the rumen fungi to higher population densities would be beneficial in increasing the nitrogen flow to the animal.

This information is somewhat at variance with what we know of the survival of fungi in the lower alimentary tract of ruminants (75). The survival of some rumen fungi in the alimentary tract after flowing out of the rumen indicates either that a stage resistant to digestion is developed or that the fungi can endure partial digestion and remain viable. Transit through the omasum may in some way render the rumen fungi less sensitive to digestion or trigger the development of a resistant stage. Microscopic examination of gut contents has not revealed motile zoospores in any part of the tract other than the rumen, and vegetative growth is observed only occasionally (C. G. Orpin, unpublished).

Anaerobic fungi from the hindgut-fermenting herbivores must represent a loss of both protein and carbon to the animal, since the vegetative growth is voided in the feces (25,36). However, the ability of the fungi to hydrolyze plant cell wall carbohydrates could offset this loss to the animal.

The rumen fungi *Neocallimastix* spp. can take up and incorporate lysine, tyrosine and methionine unchanged into cellular protein (92). The addition of a range of amino acids to the defined minimal medium used to grow *N. patriciarum* resulted in increased growth (94), and the provision of ammonium ions and L-cysteine was sufficient to allow growth in the same species, suggesting that of all the required cellular amino acids could be synthesized from these two compounds.

VII. BIOCHEMISTRY

A. Fermentation Products

All the strains of anaerobic fungi so far examined produce a mixed acid fermentation from carbohydrates, with the production of principally formic, acetic, and lactic acids, ethanol, hydrogen, and carbon dioxide (19,95,96,97). The lactic acid is chiefly of the D(−) isomer, though traces of L(+)-lactate have been detected in culture supernatant fluids of some strains (95). Succinic acid has also been found to be produced in significant quantitites by some strains (C. G. Orpin, unpublished), particularly during the latter phases of the growth cycle.

The pattern of fermentation products is markedly affected by the presence of methanogenic bacteria (19,98,99) that can utilize the hydrogen and formate generated by the fungus. This removal of fermentation products results in more rapid hydrolysis of the substrate and growth of the fungus (19,99) and is probably more representative of the situation in the rumen than that shown by pure cultures of fungi grown in vitro.

Anaerobic fungi can utilize a wide range of mono-, di-, and polysaccharides for growth, including the plant polysaccharides cellulose, hemicelluloses, and starch (18,19,20,25,95,101). Pectin is not normally fermented, but there is one report of an isolate that can utilize pectin (100). It is interesting to note that rumen fungi can colonize the pre-rumen, before the animal starts to eat plant material (77). Most strains of anaerobic fungi will grow well on lactose (C. G. Orpin, unpublished), which is present in milk and probably enters the pre-rumen in small quantities.

B. Extracellular Proteins

A wide range of enzyme activities involved in plant cell-wall hydrolysis are released into the culture medium by growing cultures of anaerobic fungi, including cellulase, xylanase, and other hemicellulases and glycosidases (21, 22,23,25,100–103). In forage grasses the dominant cell wall polysaccharides are cellulose, glucoarabinoxylans, and varying quantities of $(1 \rightarrow 3, 1 \rightarrow 4)$-$\beta$-glucans (104). Dicots, which also feature as forage plants such as legumes, contain, in addition to cellulose, pectic polysaccharides, xyloglucans, and other heteroxylans (104).

Enzymes involved in cellulolysis are endo-1,4-β glucanase, cellobiohydrolase, and β-glucosidases, and these enzyme activities are present in the supernatant fluid of cultures of *Neocallimastix* spp. (21,22,23,96,100), *Piromyces* spp. (21,22,23), *Sphaeromonas* (21), and *Orpinomyces bovis* (101). In *Neo-*

callimastix patriciarum and *Piromyces* spp., cellulase and xylanase activities were produced whether the fungi were grown on cellulose or on glucose (25). Activities in culture supernatant fluids grown on cellulose were, however, considerably higher than in those grown on glucose, suggesting a low level of constitutive expression as well as induction in the presence of cellulose. This is supported by the discovery of at least three cellulase genes of *N. patriciarum* that are induced when the organism is grown on cellulose (105), and a fourth (*celD*) that is expressed constitutively (106). The *celD* enzyme is multifunctional and encodes endo- and exo-1,4-β-glucanase and xylanase activities (105). Xylosidase, arabinosidase, glucosidase, and xylanase activities are also found in culture supernatant fluids (21,22,23,96,102,103). Pectin is released from plant tissues during growth of the rumen fungi, but only one strain (100) has been shown to use pectin as growth carbohydrate. Low levels of pectin-hydrolyzing enzymes are, however, produced by some strains (23). Thus, with the possible exception of pectin, the major plant cell wall poly-saccharides may be hydrolyzed by enzymes produced by the rumen fungi, and the component sugars used for growth. Other plant polysaccharides that support the growth of anaerobic fungi include laminarin, barley β-glucan, and pustulan (107).

During growth on wheat straw and grass hay, rumen fungi release lignin from the substrate. This is released partly as lignin–carbohydrate moieties (108) and partly by release of simple aromatic structures such as ferulic and *p*-coumaric acid (101,109). Feruloyl-, *p*-coumaroyl-, and acetyl-esterase activities have been detected in culture supernatants of *Piromyces* spp., *Neocallimastix* spp., and *Orpinomyces* spp., and a proteolytic uniflagellate species (101). Both feruloyl- and *p*-coumaroyl-esterases have been purified (110, 111) from *Neocallimastix* strain MC-2.

Protease is also secreted into culture supernatants by *Neocallimastix frontalis* (112), which may affect the activities of secreted plant cell-wall–degrading enzymes.

C. Proteolysis

A strain of *Neocallimastix frontalis* isolated from sheep was shown to possess high protease activity that was associated both with the fungal cells and with culture supernatant fluids. The protease had a pH optimum of 7.5 and was tentatively identified as a metalloprotease of activity comparable to that of the most proteolytic rumen bacteria and subject to a complicated regulatory system that was not elucidated (112). The function of the protease has yet to be determined, but it may be involved in hydrolyzing structural protein in plant cell walls to aid penetration by the rhizoids, or it could be used for the generation of amino acids for growth. It may also modify the structure or activities of other extracellular enzymes involved in plant cell wall digestion by the fungus. Possible support for the latter theory is provided by the discovery of seven different cellobiohydrolases and three endo-1,4-β-glucanases in culture supernatants of rumen fungi (113). Some of these may have been produced by activity of the protease, but until these cellulases and the genes coding for their synthesis have been examined in detail it is impossible to determine the role, if any, of the protease in the generation of multiple cellulases. It was estimated that *N. frontalis* had the potential to contribute up to 50% of the total proteolytic activity in the rumen (112).

D. Lactate Dehydrogenase

D($-$)-lactate dehydrogenase [D($-$)-lactate:NAD oxidoreductase, E.C.1.1.1.28] occurs in only a few species of Chytridiomycetes and in the anaerobic rumen fungi *Neocallimastix* spp., *Piromyces*, and *Sphaeromonas* (114), supporting a phylogenetic relatedness between the anaerobic fungi and the other Chytridiomycetes. The other Chytridiomycetes possessing D($-$)-lactate dehydrogenase are *Allomyces*, *Blastocladiella*, and *Blastocladia* and some Oomycetes (115,116,117).

E. Effects of Growth Promoters and Antibiotics

The ionophores and other antibiotics are often use in the cattle industry to improve the feeding efficiency of growing animals and to treat digestive disorders. In pure culture in the laboratory, rumen fungi are very sensitive to inhibition by ionophores (118,119,120), and it has been claimed that treatment with ionophore antibiotics may eliminate anaerobic fungi from the rumen (58). While monensin can be used to substantially reduce the number of the fungi and apparently eliminate them from the rumen for short periods, some strains apparently survive at low population densities and eventually grow and repopulate the rumen, even in the presence of the antibiotic (C. G. Orpin and L. Blackall, unpublished). When one attempts to remove rumen fungi using antibiotics, their survival in saliva and feces should be borne in mind; it suggests that stringent precautions should be taken to eliminate reinoculation from the feces of the same animal and the saliva and feces of other animals.

Rumen anaerobic fungi were found to be resistant to some other antibiotics, including avoparcin, bacitracin, tylosin, and virginiomycin (C. G. Orpin, unpublished observations). However, removal from the rumen by treatment with cycloheximide, followed by isolation and the introduction of a pure laboratory culture, was sucessful in establishing a monoculture of *Neocallimastix patriciarum* in the rumen of sheep (9). Further work using this method showed that treatment with cycloheximide alone is not always successful in eliminating rumen fungi, and it often results in the treated animal ceasing to eat (C. G. Orpin, unpublished observations).

VIII. CONCLUSIONS

Although much is now known conerning the general biology, chemistry, and biochemistry of anaerobic fungi from the rumen, considerable gaps still exit in our knowledge. It is clear that the life cycles are more complex than originally described and contain an oxygen-resistant survival stage, and that the rumen cycle may involve the development of spores other than zoospores. Whether these two stages are the same remains to be seen. In addition, it is not known whether any sexual cycles occur.

We know little about the anaerobic fungi from the hindgut of nonruminant herbivores. Morphologically they are similar to some genera occurring in the rumen. However, it is not known whether some of these are the same species as occur in the rumen or whether there are any species-specific anaerobic fungi. Since no *Neocallimastix* spp. have been isolated from hindgut fermenters, there are clearly some differences between the populations.

Further work is necessary to define fully the taxonomic position of these unique organisms among the fungi. Current evidence suggests that they are closely related to the Chytridiomycetes, but significant structural differences at the fine-structural level (discussed in Chapter 2), the occurrence of exogenously and endogenously developing genera, and the existence of multiflagellated zoospores, unique among zoosporic fungi, all contribute to the suggestion that the anaerobic fungi may represent a distinct taxon.

The biochemistry of these fungi is regulated to a large extent by their anaerobic nature, which is clearly exemplified by their unusual lipid metabolism and composition. In addition, although these organisms are known to actively ferment plant structural carbohydrates with the release of phenolic compounds, the extent to which they are able to achieve plant cell-wall metabolism is limited, as for all rumen organisms, by the absence of molecular oxygen for the metabolism of the aromatic moieties in lignin. However, their role in hydrolyzing lignin–carbohydrate bonds may be crucial to effective ruminal digestion of plant fiber.

REFERENCES

1. Hungate RE. The Rumen and Its Microbes. London: Academic Press, 1965.
2. Bauchop T, Clarke RT. Microbial Ecology of the Gut. London: Academic Press, 1977.
3. Hobson PN. The Rumen Microbial Ecosystem. London: Elsevier Applied Science, 1988.
4. Liebetanz E. Die parasitischen Protozoen der Wiederkauermagens. Arch Prot 1910; 19:19–90.
5. Braune R. Untersuchungen uber die in Wiederkauermagen vorkommenden Protozoen. Arch Prot 1913; 32:111–170.

6. Hsuing TS. A monograph on the protozoa of the large intestine of the Horse. Iowa State Coll J Sci 1929; 4:359-343.

7. Bovee EC. Inquilinic protozoa from freshwater gastropods. II. *Callimastix jo-lepsi* n.sp. from the intestine of the pulmonate freshwater snail, *Helisoma duryi* Say, in Florida. Quart J Florida Acad Sci 1961; 24:208-214.

8. Vavra J, Joyon L. Etude sur la morphologie, le cycle évolutif at la position systématique de *Callimastix cyclopis* Weissenberg 1912. Protistologica 1966; 2: 5-15.

9. Orpin CG. Studies on the rumen flagellate *Neocallimastix frontalis*. J. Gen Microbiol 1975; 91:249-262.

10. Jensen EHC, Hammond DM. A morphological study of trichomonads and related flagellates from the bovine digestive tract. J Protozool 1964; 11:386-394.

11. Orpin CG. The occurrence of chitin in the cell walls of the rumen organisms *Neocallimstix frontalis*, *Piromonas communis* and *Sphaeromonas communis*. J Gen Microbiol 1977; 99:215-218.

12. Bauchop T. Rumen anaerobic fungi of cattle and sheep. Appl Environ Microbiol 1979; 38:148-158.

13. Orpin CG. Invasion of plant tissue in the rumen by the flagellate *Neocallimastix frontalis*. J Gen Microbiol 1977; 98:423-430.

14. Orpin CG. The rumen flagellate *Piromonas communis*: its life-history and invasion of plant material in the rumen. J Gen Microbiol 1977; 99:107-117.

15. Bauchop T. Scanning electron microscopy in the study of microbial digestion of plant fragments in the gut. In: Ellwood DC, Hedger JN, Latham MJ, et al., eds. Contemporary Microbial Ecology. London: Academic Press, 1980:305-326.

16. Orpin CG. studies on the rumen flagellate *Sphaeromonas communis*. J Gen Microbiol 1976; 94:270-280.

17. Theodorou MK, Trinci APJ. Procedures for the isolation of rumen fungi. In: Nolan JV, Leng RA, Demeyer DI, eds: The Roles of Protozoa and Fungi in Ruminant Digestion. Armidale, Australia: Penamul Books, 1989:145-152.

18. Orpin CG, Letcher AJ. utilization of cellulose, starch, xylan and other hemicelluloses for growth by the rumen phycomycete *Neocallimastix frontalis*. Curr Microbiol 1979; 3:121-124.

19. Bauchop T, Mountfort DO. Cellulose fermentation by a rumen anaerobic fungus in both the absence and presence of rumen methanogens. Appl Environ Microbiol 1981; 42:1103-1110.

20. Lowe SE, Theodorou MK, Trinci APJ. Growth and fermentation of an anaerobic rumen fungus on various carbon sources and effect of temperature on development. Appl Environ Microbiol 1987; 53:1210-1223.

21. Hebraud M, Fevre M. Characterisation of glycoside and polysaccharide hydrolases secreted by the rumen anaerobic fungi *Neocallimastix frontalis*, *Sphaeromonas communis* and *Piromonas communis*. J Gen Microbiol 1988; 134:1123-1129.

22. Williams AG, Orpin CG. Glycoside hydrolase enzymes present in the zoospore and vegetative stages of the rumen fungi *Neocallimastix patriciarum*, *Piromonas communis* and an unidentified isolate grown on a variety of carbohydrates. Can J Bot 1987; 33:427-434.

23. Williams AG, Orpin CG. Polysaccharide degrading enzymes formed by three species of anaerobic fungi grown on a arange of carbohydrate substrates. Can J Bot 1987; 33:418–426.
24. Theodorou MK, Gill M, King-Spooner C, Beever DE. Enumeration of anaerobic Chytridiomycetes as thallus-forming units: novel method for quantification of fibrolytic fungal populations from the digestive tract ecosystem. Appl Environ Microbiol 1990; 56:1073–1078.
25. Orpin CG. Isolation of cellulolytic phycomycete fungi from the caecum of the horse. J Gen Microbiol 1961; 123:287–296.
26. Whisler HC, Zebold SL, Shemanchuk JA. Alternate host for mosquito parasite *Coelomyces*. Nature 1974; 251:715–716.
27. Orpin CG, Joblin KN. Anaerobic fungi. In Hobson PN, ed. The Rumen Microbial Ecosystem. London: Elsevier Applied Science, 1988:129–150.
28. Bauchop T. Biology of gut anaerobic fungi. BioSystems 1989; 23:53–64.
29. Sparrow FK. Aquatic Phycomycetes. 2nd ed. Ann Arbor, Michigan: University of Michigan, 1973.
30. Barr DJS. How modern systematics relates to the rumen fungi. BioSystems 1988; 21:351–356.
31. Barr DJS. An outline for the reclassification of the Chytridiales and for a new order, the Spizellomycetales. Can J Bot 1980; 58:2380–2394.
32. Heath IB, Bauchop T, Skipp RA. Assignment of the rumen anaerobe *Neocallimastix frontalis* to the Spizellomycetales (Chytridiomycetes) on the basis of its polyflagellate zoospore ultrastructure. Can J Bot 1983; 61:295–307.
33. Barr DJS, Kudo H, Jackober KD, Cheng K-J. Morphology and development of rumen fungi: *Neocallimastix* sp., *Piromyces communis* and *Orpinomyces bovis*, gen. nov., sp. nov. Can J Bot 1989; 67:2815–2824.
34. Heath IB. Recommendations for future taxonomic studies of gut fungi. BioSystems 1988; 21:417–418.
35. Munn EA, Orpin CG, Hall FJ. Ultrastructural studies of the free zoospore of the rumen phycomycete *Neocallimastix frontalis*. J Gen Microbiol 1981; 125:311–323.
36. Gold JJ, Heath IB, Bauchop T. Ultrastructural description of a new chytrid genus of caecum anaerobe, *Caecomyces equi* gen. nov. BioSystems 1988; 21:403–415.
37. Munn EA, Orpin CG, Greenwood CA. The ultrastructure and possible relationships of four obligate anaerobic chytridiomycete fungi from the rumen of sheep. BioSystems 1988; 22:67–81.
38. Li J, Heath IB, Bauchop T. *Piromyces mae* and *Piromyces dumbonica*, two new species of uniflagellate anaerobic chytridiomycete fungi from the hindgut of the the horse and elephant. Can J Bot 1990; 68:1021–1033.
39. Webb J, Theodorou MK. *Neocallimastix hurleyensis* sp. nov. A new species of the genus *Neocallimastix*. Can J Bot 1991; 69:1220–1224.
40. Heath IB, Bauchop T. Mitosis and the phylogeny of the genus *Neocallimastix*. Can J Bot 1985; 63:1595–1604.

41. Li J, Heath IB, Packer L. The phylogenetic relationships of the anaerobic chytridiomycetous gut fungi (Neocallimasticaceae) and the Chytridiomycota. II. Cladistic analysis of structural data and description of the Neocallimasticales ord. nov. Can J Bot 1993; 71:393–407.
42. Li J, Heath IB. The phylogenetic relationships of the anaerobic chytridiomycetous gut fungi (Neocallimasticaceae) and the Chytridiomycota. I. Cladistic analysis of rRNA sequences. Can J Bot 1992; 70:1738–1746.
43. Kudo H, Jakober KD, Phillipe RC, et al. Isolation and characterization of cellulolytic anaerobic chytridiomycete fungi from the rumen of sheep. Can J Microbiol 1990; 36:513–517.
44. Orpin CG, Munn EA. *Neocallimastix patriciarum*: A new member of the Neocallimasticaceae inhabiting the sheep rumen. Trans Br Mycol Soc 1986; 86:103–109.
45. Webb J, Theodorou MK. A rumen anaerobic fungus of the genus *Neocallimastix*: Ultrastructure of the polyflagellate zoospore and young thallus. BioSystems 1988; 21:393–401.
46. Munn EA, Greenwood CA, Orpin CG. Organisation of the kinetosomes and associated structures of the zoospores of the rumen Chytridiomycete *Neocallimastix*. Can J Bot 1987; 65:456–465.
47. Yarlett N, Orpin CG, Munn EA, et al. Evidence for hydrogenosomes in the rumen fungus *Neocallimastix patriciarum*. biochem J 1986; 236:729–739.
48. Breton A, Bernalier A, Bonnemoy F, et al. Morphology and metabolic characterization of a new species of strictly anaerobic rumen fungus: *Neocallimastix joyonii*. FEMS Microbiol Lett 1989; 58:309–314.
49. Wubah DA, Fuller MS. Studies on *Caecomyces communis*: Morphology and development. Mycologia 1991; 83:303–310.
50. Ho YW, Bauchop T. *Ruminomyces elegans* gen. et sp. nov. A polycentric anaerobic rumen fungus from cattle. Mycotaxon 1990; 38:397–405.
51. Akin DE, Borneman WS, Windham WR. Rumen fungi: Morphological types from three Georgia cattle and the attack on forage cell walls. BioSystems 1988; 21:385–391.
52. Phillips MW. Unusual rumen fungi isolated from northern Australian cattle and water buffalo. In: Nolan JV, Leng RA, Demeyer DI, eds. The Roles of Protozoa and Fungi in Ruminant Nutrition. Armidale, Australia: Penambul Books, 1988:247–249.
53. Breton A, Bernalier A, Dusser M, et al. *Anaeromyces mucronatus* nov. gen., nov. sp. A new strictly anaerobic rumen fungus with polycentric thallus. FEMS Microbiol Lett 1990; 70:177–182.
54. Orpin CG. The characterisation of the rumen bacterium Eadie's Oval, *Magnoovum* gen. nov. *eadii* sp. nov. Arch Microbiol 1976; 111:155–159.
55. Orpin CG. The culture *in vitro* of the rumen bacterium Quin's Oval. J Gen Microbiol 1972; 73:523–530.
56. Prins RA. Isolation, culture and fermentation characteristics of *Selenomonas ruminantium* var *bryanti* var. n. from the rumen of sheep. J Bacteriol 1971; 105: 820–825.

57. Coleman GS. Rumen entodiniomorphid protozoa. In: Taylor AER, Baker JR, eds. Methods of Cultivating Parasites *in vitro*. London: Academic Press, 1978;

58. Elliott R, Ash AJ, Calderton-Cortes F, et al. The influence of anaerobic fungi on rumen volatile fatty acid concentrations in vivo. J Agri Sci Cambr 1987; 109: 13–17.

59. Orpin CG. The rumen flagellate *Callimastix frontalis*: Does sequestration occur? J Gen Microbiol 1974; 84:395–398.

60. Joblin KN. Isolation, enumeration and maintenance of rumen anaerobic fungi in roll tubes. Appl Environ Microbiol 1981; 42:1119–1122.

61. Ogimoto K, Imai S. Atlas of Rumen Microbiology. Tokyo: Japan Scientific Societies Press, 1981.

62. Milne A, Theodorou MK, Jordan MGC, et al. Survival of anaerobic fungi in faeces, in saliva and in pure culture. Exp Mycol 1989; 13:27–37.

63. Bauchop T. The gut anaerobic fungi: colonisers of dietary fibre. In: Wallace G, Bell L, eds. Fibre in Human and Animal Nutrition. Wellington, New Zealand: Royal Society of New Zealand, 1953:143–148.

64. Orpin CG, mathiesen SD, Greenwood Y, Blix AS. Seasonal changes in the rumen microflora of the Svalbard reindeer (*Rangifer tarandus platyrhynchus*). Appl Environ Microbiol 1986; 50:144–151.

65. Teunissen MJ, Op den Camp HJM, Orpin CG, et al. Growth characteristics of anaerobic fungi isolated from several herbivorous mammals during cultivation in a novel defined medium. J Gen Microbiol 1991; 137:1401–1408.

66. Warner ACS. Diurnal changes in the concentrations of microorganisms in the rumens of sheep fed a limited diet once daily. J Gen Microbiol 1966; 45:213–235.

67. Orpin CG. On the induction of zoosporogenesis in the rumen phycomycetes *Neocallimastix frontalis*, *Piromonas communis* and *Sphaeromonas communis*. J Gen Microbiol 1977; 101:181–189.

68. Orpin CG. The effects of haems and related compounds on growth and zoosporogenesis of the rumen phycomycete *Neocallimastix frontalis* H8. J Gen Microbiol 1986; 132:2179–2185.

69. Lowe SE, Griffith GW, Milne A, et al. The life-cycle and growth kinetics of an anaerobic rumen fungus. J Gen Microbiol 1987; 133:1815–1827.

70. Orpin CG, Bountiff L. Zoospore chemotaxis in the rumen phycomycete *Neocallimastix frontalis*. J Gen Mcrobiol 1978; 104: 113–122.

71. Joblin KN. Physical disruption of plant fibre by rumen fungi of the *Sphaeromonas* group. In: Nolan JV, Leng RA, Demeyer DI, eds. The Roles of Protozoa and Fungi in Ruminant Digestion. Armidale, Australia: Penambul Books, 1989:259–260.

72. Ho YW, Abdullah N, Jalaludin S. Penetrating structures of anaerobic rumen fungi in cattle and swamp buffalo. J Gen Microbiol 1988; 134:177–182.

73. Breton A, Galliard B, Fonty G. Somatic structures of the anaerobic rumen fungi. In: Nolan JV, Leng RA, Demeyer DI, eds. The Roles of Protozoa and Fungi in Ruminant Digestion. Armidale, Australia: Penambul Books, 1989:261–262.

74. Galliard B, Breton A, Bernalier A. Study of the nuclear cycle of four species of strictly nanaerobic rumen fungi by fluorescence microscopy. Curr Microbiol 1989; 13:103–107.

75. Theodorou MK, Davies D, Jordan MGC, et al. Origin and characteristics of anaerobic fungi in faeces. Exp Mycol. In press.
76. Lowe SE, Theodorou MK, Trinci APJ. Isolation of anaerobic rumen fungi from saliva and faeces of sheep. j Gen Microbiol 1987; 133:1829-1834.
77. Fonty G, Gouet PH, jouanny JP, Senaud J. Establishment of the microflora and anaerobic fungi in the rumen of lambs. J Gen Microbiol 1987; 133:1835-1843.
78. Orpin CG. Ecology of rumen anaerobic fungi in relation to the nutrition of the host animal. In: Nolan JV, Leng RA, Demeyer DI, eds. The Roles of Protozoa and Fungi in Ruminant Digestion. Armidale, Australia: Penambul Books, 1989: 29-38.
79. Phillips MW, Gordon GLR. Growth characteristics on cellobiose of three different anaerobic fungi isolated from the ovine rumen. Appl Environ Microbiol 1989; 55:1695-1702.
80. Patton RS, Chandler PT. In vivo digestibility evaluation of chitinous materials. J Dairy Sci 1975; 58:1945-1958.
81. Kemp P, Jordan DJ, Orpin CG. The free- and protein-amino acids of the rumen phycomycete fungi *Neocallimastix frontalis* and *Piromonas communis*. J Agri Sci Cambr 1985; 105:523-526.
82. Akin DE. Use of chitinase to determine rumen fungi with plant tissues *in vitro*. Appl Environ Microbiol 1987; 53:1955-1958.
83. Argyle JL, Douglas L. Chitin as fungal marker. In: Nolan JV, Leng RA, Demeyer DI, eds. The Roles of Protozoa and Fungi in Rumen and Digestion. Armidale, Australia: Penambul Books, 1989:289-290.
84. Gay L, Hebraud M, Girard V, Fevre M. Chitin synthase activity from *Neocallimastix frontalis*, an anaerobic rumen fungus. J Gen Microbiol 1989; 135:279-283.
85. Pfyffer GE, Pfyffer BU, Rast DM. The polyol pattern, chemotaxonomy and phylogeny of the fungi. Sydowia 1986; 39:160-202.
86. Rast DM, Pfyffer GE. Acyclic polyols and higher taxa of fungi. Bot J Linn Soc 1989; 99:39-57.
87. Pfyffer GE, Boraschi-Gaia C, Weber B, et al. A further report on the occurrence of acyclic alcohols in fungi. Mycol Res 1990; 94:219-222.
88. Pfyffer GE, Rast DM. The polyol pattern of some fungi not hitherto investigated for sugar alcohols. Exp Mycol 1980; 4:160-170.
89. Kemp P, Lander D, Orpin CG. The lipids of the anaerobic rumen fungus *Piromonas communis*. J. Gen Microbiol 1984; 130:27-37.
90. Body DR, Bauchop T. Lipid composition of the obligately anaerobic fungus *Neocallimastix frontalis* isolated from a bovine rumen. Can J Microbiol 1985; 31:463-466.
91. Tchen TT, Bloch K. On the mechanism of enzymic cyclisation of squalene. J Biol Chem 1957; 226:931-938.
92. Gulati SK, Ashes JR, Gordon GLR, et al. Nutritional availability of amino acids from the rumen anaerobic fungus *Neocallimastix* sp. LM1 in sheep. J Agri Sci Cambr 1989; 113:383-387.

93. Elliott R, Little D. The true absorption of cyst(e)ine from the ovine small intestine. Br J Nutr 1977; 37:285–287.

94. Orpin CG, Greenwood Y. Nutritional and germination requirements of the rumen phycomycete *Neocallimastix patriciarum*. Trans Br Mycol Soc 1986; 86:178–181.

95. Phillips MW, Gordon GLR. Sugar and polysaccharide fermentation by rumen anaerobic fungi from Australia, Britain and New Zealand. BioSystems 1988; 21:377–383.

96. Richardson AJ, Calder AG, Stewart CS, Smith A. Simultaneous determination of volatile and non-volatile acidic fermentation products of anaerobes by capillary gas chromatography. Lett Appl Microbiol 1989; 9:5–8.

97. Gordon GLR, Phillips MW. Comparative fermentation properties of anaerobic fungi from the rumen. In: Nolan JV, Leng RE, Demeyer DI, eds. The Roles of Protozoa and Fungi in Ruminant Digestion. Armidale, Australia: Penambul Books, 1989:127–138.

98. Joblin KN, Naylor G, Williams AG. The effect of *Methanobrevibacter smithii* on the xylanolytic activity of anaerobic rumen fungi. Appl Environ Microbiol 1990; 56:2287–2295.

99. Mountfort DO, Asher RA, Bauchop T. Fermentation of cellulose to methane and carbon dioxide by a rumen anaerobic fungus in a triculture with *Methanobrevibacter smithii* sp. strain RA1 and *Methanosarcina barkeri*. Appl Environ Microbiol 1982; 44:128–134.

100. Barrievich EM, Calza RE. Media carbon induction of extracellular cellulase activities in *Neocallimastix frontalis* EB188. Curr Microbiol 1990; 20:265–271.

101. Bornemann WS, Akin DE. Lignocellulose degradation by rumen fungi and bacteria: ultrastructure and cell-wall degrading enzymes. In: Akin DE, Ljungdahl LG, Wilson JR, Harris PJ, eds. Microbial and Plant Opportunities to Improve Lignocellulose Utilization by Ruminants. New York: Elsevier, 1990:325–339.

102. Lowe SE, Theodorou MK, Trinci APJ. Cellulases and xylanase of an anaerobic rumen fungus grown on wheat straw, wheat straw holocellulose, cellulose and xylan. Appl Environ Microbiol 1987; 53:1216–1223.

103. Hebraud M, Fevre M. Purification and characterization of an extracellular β-xylosidase from the rumen anaerobic fungus *Neocallimastix frontalis*. FEMS Microbiol Lett 1990; 72:11–16.

104. Harris PJ. Plant cell wall structure and development. In: Akin DE, Ljungdahl LG, Wilson JR, Harris PJ, eds. Microbial and Plant Opportunities to Improve Lignocellulose Utilization by Ruminants. New York: Elsevier, 1990:71–90.

105. Xue GP, Orpin CG, Gobius KS, et al. Clining and expression of multiple cellulase cDNAs from the anaerobic rumen fungus *Neocallimastix patriciarum* in *Escherichia coli*. J Gen Microbiol 1992; 138:1413–1420.

106. Xue GP, Gobius KS, Orpin CG. A novel polysaccharide hydrolase cDNA (celD) from *Neocallimastix patriciarum* encoding three multifunctional cata-

lytic domains with high endoglucanase, cellobiohydrolase and xylanase activities. J Gen Microbiol 1992; 138:2397–2403.

107. Gordon GLR. Selection of anaerobic fungi for better fiber degradation in the rumen. In: Akin DE, Ljungdahl LG, Wilson JR, Harris PJ, eds. Microbial and Plant Opportunities to Improve Lignocellulose Utilization by Ruminants. New York: Elsevier, 1990:301–310.

108. Orpin CG. The roles of ciliate protozoa and fungi in the rumen digestion of plant cell walls. Anim Freed Sci Technol 1983/4; 10:121–143.

109. Bornemann WS, Hartley RD, Morrison WH, et al. Feruloyl and *p*-coumaroyl esterase from anaerobic fungi in relation to plant cell wall degradation. Appl Microbiol Biotechnol 1990; 33:345–351.

110. Bornemann WS, Ljungdahl LG, Hartley RD, Akin DE. Isolation and characterization of of *p*-coumaroyl esterase from the rumen anaerobic fungus *Neocallimastix* strain MC-2. Appl Environ Microbiol 1991; 57:2337–2344.

111. Bornemann WS, Ljungdahl LG, Hartley RD, Akin DE. Purification and partial characterization of two feruloyl esterases from the anaerobic fungus *Neocallimastix* strain MC-2. Appl Environ Microbiol 1992; 58:3762–3766.

112. Wallace RJ, Joblin KN. Proteolytic activity of a rumen anaerobic fungus. FEMS Microbiol Lett 1985; 29:19–25.

113. Barrievitch EM, Calza RE. Supernatant protein and cellulase activities of the anaerobic rumen fungus *Neocallimastix frontalis* EB188. Appl Environ Microbiol 1990; 56:43–48.

114. Gleason FH, Gordon GLR. Lactate dehydrogenases in obligately anaerobic Chytridiomycetes from the rumen. Mycologia 1990; 82:261–263.

115. Purrohit K, Turian G. D(–)lactate dehydrogenases from Allomyces. Partial purification and allosteric properties. Arch Mikrobiol 1972; 84:287–300.

116. Rivedal E, Sanner T. Properties of D(–)lactate dehydrogenase from *Blastocladiella emersonii*. Exp Mycol 1978; 2:337–345.

117. Gleason FH, Price JS. Lactic acid fermentation in the lower fungi. Mycologia 1969; 61:945–956.

118. Marounek M, Hodrova B. Susceptibility and resistance of anaerobic rumen fungi to antimicrobial feed additives. Lett Appl Microbiol 1989; 9:173–175.

119. Stewart CS, McPherson CA, Cansunar E. The effect of lasalocid on glucose uptake, hydrogen production and the solubilization of straw by anaerobic rumen fungus *Neocallimastix frontalis*. Lett Appl Microbiol 1987; 5:5–7.

120. Stewart CS, Richardson AJ. Enhanced resistance of anaerobic rumen fungi to the ionophores monensin and lasalocid in the presence of methanogenic bacteria. Lett Appl Microbiol 1989; 6:85–93.

121. Wubah DA, Fuller MS, Akin DE. Resistant body formation in *Neocallimastix* sp. an anaerobic fungus from the rumen of cow. Mycologia 1991; 83:40–47.

122. Ho YW, Abdullah N, Jalaludin S. Colonization of guinea grass by anaerobic rumen fungi in swamp buffalo and cattle. Anim Feed Sci Technol 1988; 22:161–172.

2

The Ultrastructure of Anaerobic Fungi

EDWARD A. MUNN

The Babraham Institute
Babraham, Cambridge, England

I. INTRODUCTION

For the purposes of this chapter ultrastructure is taken to mean structure revealed by electron microscopy. Three basic techniques, thin-sectioning and negative-staining transmission electron microscopy and scanning electron microscopy, have been applied to the study of anaerobic fungi. Most of the data presented in this chapter will describe results obtained using thin sectioning, supplemented with information from negative staining where available. Scanning electron microscopy as so far applied in this area has provided useful information on the surface structure of vegetative growth stages in particular, and it is the only electron microscopy technique that has been used extensively with samples taken directly from the digestive tract (1,2,3). Nearly all transmission electron microscopy, i.e., that which has yielded information about the internal structure of the fungi, has been applied to isolates cultured in vitro. Many individuals from these cultures may be studied to yield the published information, but the sample studied is small at best. In no case has the extent to which it is representative of the population from which it was isolated been examined. It follows that where structural similarities are reported between isolates, especially when these are from different laboratories or from different hosts, their validity for the population as a whole is likely to be high. Where some subtle difference in structure between isolates is described, the validity of this for the population as a whole is less certain.

In the absence of evidence to the contrary, it may be assumed that the possession of some distinct ultrastructural feature is a reflection of the possession of some distinct functional feature; often, however, a structure will be described without the ascription of a function.

II. LIFE CYCLE

The life cycle is shown diagramatically in Fig. 1,* in which some ultrastructural features are presented. These are considered in detail in the following sections. Although all stages of the life cycle have been examined by electron microscopy, most attention has been given to the free zoospores, i.e., zoospores that have completed differentiation and been released from the sporangia.

A. Free Zoospores

Descriptions of the ultrastructural features of free zoospores of three species of *Neocallimastix* from the sheep rumen, *N. patriciarum* (originally described as *N. frontalis* [4–7], *N. frontalis* (PN1) (8), and *N. hurleyensis* (9,10), have been published. Additional, limited data on a *Neocallimastix* (NRE10) from the rumen of Svalbard reindeer and a *Neocallimastix* (CM1) from the rumen of Northern Territories cattle are fully compatible with these descriptions. There are also descriptions of *Piromyces communis, P. lethargica* (NF1), and *Sphaeromonas communis* from the sheep rumen (6); *Caecomyces equi* (11) and *Piromyces mae* (12) from horse dung and *P. dumbonica* (12) from elephant dung, all three cultures unfortunately no longer extant; and *Piromyces equi* from the horse cecum (13).

Although individual ultrastructural features have been used to distinguish these genera and species, overall they have many features in common. They share some of these features — for example, a spiral (14), also called a cylinder (8), concentric with the basal end of the central pair of microtubules of each flagellum and a spur (15) adjacent to each kinetosome — with aerobic members of the Chytridiomycetes, and indeed, this is the ultrastructural basis for including them in this taxon (8). Other ultrastructural features collectively, or some even individually, distinguish them from the zoospores of aerobic fungi (6). In particular there are hydrogenosomes, but no mitochondria. There are two kinds of aggregate of ribosomelike particles, helices and

---

*See Abbreviations at end of chapter.

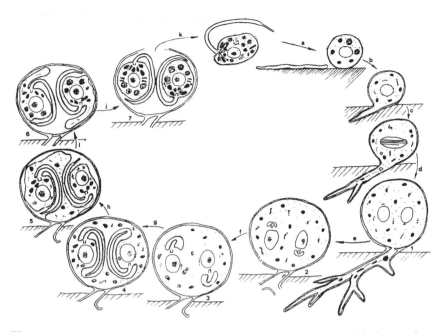

Figure 1 Diagram of the typical life cycle of a monocentric anaerobic fungus showing the stages that have been examined by electron microscopy. (a) The free zoospore is attracted chemotactically to fresh plant material, to which it attaches, usually losing its flagellum; (b) the encysted zoospore develops a rhizoid and (c) the vegetative growth phase with extensive rhizoid system becomes established; nuclear division begins. (d) Mitotic divisions produce two or more nuclei, and a cross-wall forms to separate the developing sporangium (1) from the rhizoid system. (e) The precursors of the flagellar cleavage vesicles appear adjacent to the kinetosomes; (f) The cleavage vesicles begin to enlarge and the flagella to elongate (3). (g) This process continues, and layers begin to polymerize on what will become the plasma membrane surface. (h) The zoospores begin to detach from the sporangium wall, and the flagellar vesicles fuse with the peripheral plasma membrane (5). (i) The organelles begin to be localized in the positions they will occupy in the free zoospore. (j) Zoospore differentiation is completed (7), the sporangium wall is digested, and (k) the zoospores swim free. This diagram is slightly modified from that published by Munn et al. (6). It is based largely on studies of *Neocallimastix patriciarum*, but for clarity only one flagellum is shown per zoospore, and the diagram can therefore equally well represent the life cycle of uniflagellate anaerobic zoospores with filamentous rhizoids. All the information available indicates that the ultrastructural changes during the life cycles of the monoflagellates are essentially the same as those of *Neocallimastix*.

globules; and two distinct layers coating the flagellum and zoospore cell body plasma membrane, respectively. The flagella have narrowed tips into which only a few microtubules extend. There may be many kinetosomes, as in the *Neocallimastix* species, but there are never nonfunctional kinetosomes. Props (links from the distal ends of the triplets to the adjacent plasma membrane), typically present in zoospores of aerobic chytridiomycetes, have been looked for but, with one exception, have not been reported. There are, however, elaborate perikinetosomal structures — struts, skirt, scoop, and so on, described in detail later — that may serve the function of props. Also present are crystallites, membrane stacks, and abundant glycogen. The organization of all these features is shown in the diagram in Fig. 2, although not all the features labeled will be evident in single sections, even those in the anterior-posterior plane (Fig. 3).

1. Flagella

The flagella range in length from 24 μm for the uniflagellate *Piromyces* and *Sphaeromonas* to 37 μm for the multiflagellate zoospores of *Neocallimastix*

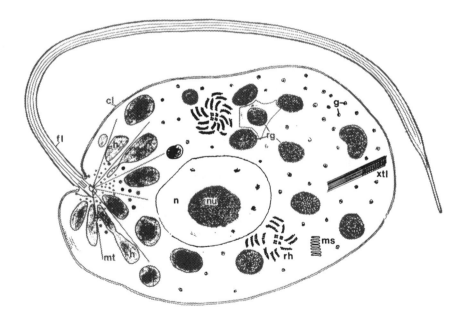

Figure 2   Composite diagram to show the main ultrastructural features of free zoospores. Not all the features labeled will be evident in single sections, even those in the anterior–posterior plane depicted (compare Fig. 3).

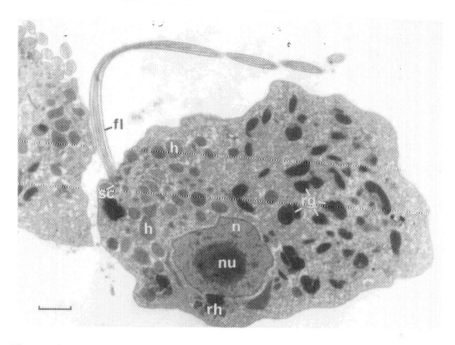

Figure 3 Electron micrograph of free zoospores of *Neocallimastix patriciarum* for comparison with the diagram shown in Fig. 2. The plane of the section is approximately anterior–posterior for the zoospore shown complete. The section passes at right angles through the double row of kinetosomes so that only one is visible. Some kinetosomes sectioned along the row are visible in the zoospore shown in part on the left. Scale bar equals 1 μm.

(6). In transverse sections the flagella show through most of their length the nine outer doublets and central pair arrangement of microtubules characteristic of eucaryotic cilia and flagella (Figs. 4, 5), but this pattern is lost near the tip, where the flagellum tapers from a diameter of about 0.3 μm to a point (Fig. 6). In this narrowed region only a few single microtubules are present, and even they peter out at the very tip.

The flagella of *Neocallimastix* beat as a single unit, and it is likely from limited electron-microscopic data that the flagellar microtubules in the individual flagella are all oriented more or less in the same plane (Fig. 4e).

At the proximal end of the flagellum, in the transition zone adjacent to the kinetosome, there is a cylinder or spiral 0.12 μm in diameter lying just inside the outer doublets. In transverse sections, this is seen clearly as a circle or enneagon with a wall about 6 nm thick, i.e., slightly thinner than the wall

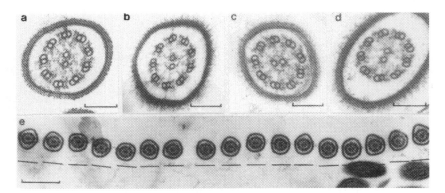

Figure 4 Electron micrographs of **(a-e)** transverse sections through flagella from **(a,e)** *N. patriciarum*; **(b)** *P. communis*; **(c)** *S. communis*; **(d)** *P. lethargica* (NF1). The group shown in **(e)** is included to show that the orientation of the individual axonemes can be very similar. Lines passing through the centers of the central pair of microtubules of each of the seven flagella in the middle of the row are parallel; the corresponding lines in the other ten flagella do not deviate by more than 8° from this plane. Scale bars equal 100 nm **(a-d)** and 0.5 μm **(e)**.

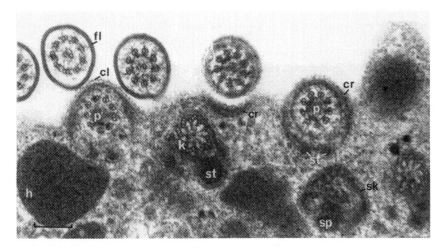

Figure 5 Electron micrograph of a transverse section through the proximal ends of some flagella and the kinetosomes adjacent to flagella of *N. patriciarum*. The section passes through these entities at various levels (compare Figs. 8–10). Scale bar equals 0.2 μm.

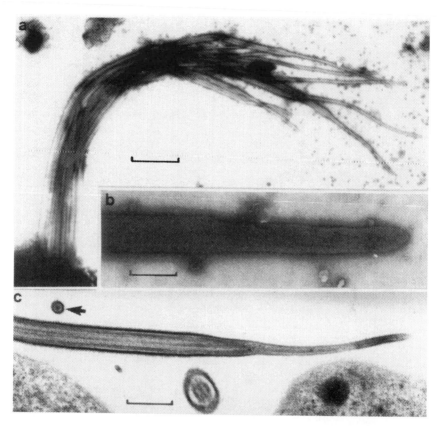

Figure 6 Electron micrographs of **(a,b)** negatively stained and **(c)** sectioned flagella. **(a)** *N. patriciarum*, **(b)** *P. communis*, and **(c)** *P. equi* showing the narrow tips to the flagella and in **(c)** in cross section (arrowed) and longitudinal section, the few microtubules contained therein. Scale bars equal 2 μm **(a,b)** and 0.5 μm **(c)**.

of the microtubules (cy, Fig. 7), but it is far less obvious in longitudinal sections (Figs. 6e, 8, 9); hence the uncertainty as to whether it is a cylinder or a spirally arranged filament as seen, for example, in the aerobic chytrids *Rhizophydium* and *Rhizophlyctis* (14). (If it is a cylinder rather than a spiral this would be of considerable phylogenetic interest.) As seen in transverse section the circle has a connection about 5 nm long to each outer doublet at its midpoint. There is a short corresponding prominence or line of some kind directed inward from the circle at the position of each connection to the doublets.

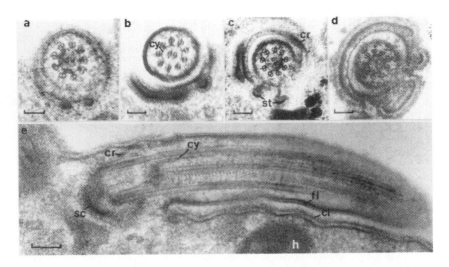

Figure 7 Electron micrographs of **(a-d)** transverse sections and **(e)** longitudinal section through the proximal ends of the flagella to show the cylinder/spiral. **(a,e)** *N. patriciarum,* **(b)** *P. communis,* **(c)** *S. communis,* **(d)** *P. equi.* Scale bars equal 0.1

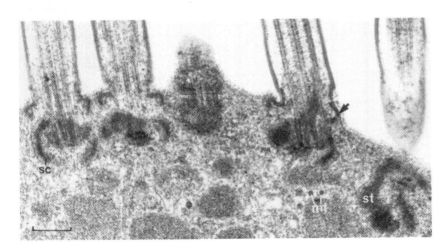

Figure 8 Electron micrograph of a longitudinal section through five kinetosomes in a free zoospore of *N. patriciarum.* The plane of sectioning of the two kinetosomes to the left and the grazing one on the right approximately parallels the plane of the central pair of flagellar microtubules and corresponds to that shown diagrammatically in Fig. 9a. The arrow indicates a structure with a superficial resemblance to a prop (see text for explanation). Scale bar equals 0.2 μm.

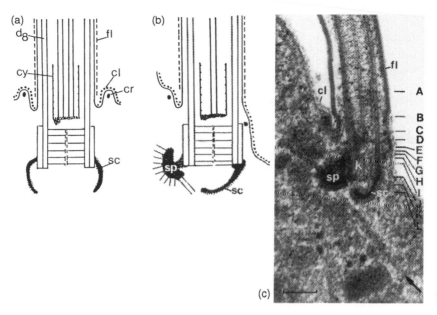

Figure 9 **(a,b)** Diagrams and **(c)** electron micrograph of longitudinal sections through kinetosomes of *N. patriciarum*. The plane of sectioning depicted in **(a)** is at right angles to that of **(b)** and **(c)**. The lines labeled A–L to the right of **(c)** indicate the approximate planes of the transverse sections shown in Fig. 10. Scale equals 0.2 mm approximately. (From Ref. 5.)

As first noted by Heath et al. (8) for *N. frontalis*, the doublets adjacent to the cylinder are connected to the flagellar plasma membrane by transition fibrils. At the proximal end of the cylinder, at least in *Neocallimastix*, there is a plate of amorphous material that has an extension into the kinetosome (see below). A transverse section passing at different levels through one or another of the proximal ends, transition zones, and kinetosomes of eight flagella of *N. patriciarum* is shown in Fig. 5.

2. Kinetosomes and Associated Structures

The difference in the number of flagella, which distinguishes the multiflagellate genus *Neocallimastix* from all other genera of anaerobic fungi, is obvious from light microscopy. What is not apparent from light microscopy but is revealed by electron microscopy is that the kinetosomes from which the flagella arise are in an ordered double-row array in all the *Neocallimastix* species (Fig. 5).

Details of the organization of the individual kinetosomes and the periki-netosomal structures have been established for both multiflagellate and uni-flagellate species. Interpretations have been based on both transverse sections (Figs. 5, 10) and longitudinal sections (Figs. 8, 9), and particular use has been made of serial sections. Interpretation of the sequences of serial transverse sections of the kinetosomes is relatively easy, because the orientation of the plane of sectioning can be verified from the clarity of the cross-sectioned microtubules and, in the case of *Neocallimastix*, because there are many kinetosomes present in any one field (Fig. 5). For longitudinal sections interpretation is more difficult, because the plane of sectioning may be parallel to any one of a range of chord angles and frequently is not strictly parallel to the longitudinal axis of the kinetosome.

The length of the kinetosome with a triplet structure is about 180 nm. As in all other kinetosomes, the triplets are inclined at an angle to the circumference of the kinetosome. At the proximal end the outer edge of the inner-most microtubule of each triplet appears to touch the inner edge of the inner-most triplet, but more distally the triplets are slightly further apart and are

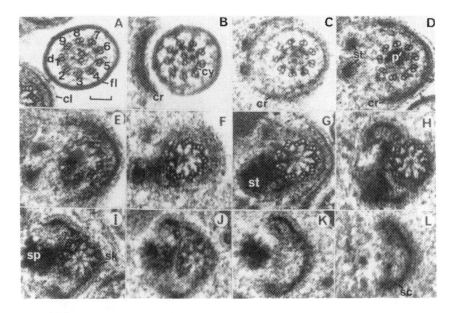

Figure 10 Electron micrographs of transverse sections of kinetosomes of *N. patriciarum* selected from serial sections similar to those shown in Fig. 5. Scale bar equals 0.1 μm. (From Ref. 5.)

connected by links like those described for aerobic fungi (14). For much of the length of each kinetosome there is an axial tubule some 15 nm in diameter, connected by radial spokes to the innermost microtubule of each triplet (see Figs. 5 and 10).

In *Neocallimastix*, the plate that marks the proximal end of the cylinder and termination of the central pair of flagellar microtubules has a proximal prolongation adjacent to doublets 8, 9, and 1 that extends into the kinetosome (Fig. 10E). Proximal to the termination of the central pair, there are no direct connections between the distal end of the triplets or the doublets and the plasma membrane (i.e., there are no props; see below).

The terminology used in describing the perikinetosomal structures — circumflagellar ring, skirt, spur, scoop, and struts — follows that of Barr (15, 16), Heath et al. (8), Li et al. (12), and Munn et al. (5,6) and is illustrated in the diagrams of longitudinal sections in Fig. 9. Collectively, these accessory structures have been called the harness (17). In stained sections, the components of the harness in general have a very electron-dense core and a less electron-dense peripheral layer up to 35 nm thick. A three-dimensional impression of these structures may be obtained from electron micrographs of a negatively stained intact kinetosome with harness, obtained from *P. equi* (Fig. 11), and photographs of a model of the harness and kinetosome of *N. patriciarum* (Fig. 12).

The circumflagellar ring (cr, Fig. 9) lies outside the outer doublets and just below the plasma membrane at the level of the amorphous plate, which marks the proximal end of the cylinder. In *N. patriciarum* this is about 400 nm in diameter and uniformly about 20 nm thick. In the *Piromyces* spp. the ring has a greater diameter (12) and appears ribbonlike, with the wider part parallel to the long axis of the flagellum. In *P. mae* the ribbon is about 20 nm thick and uniformly 70–90 nm wide, and in *P. dumbonica* the ribbon is C-shaped and 220 nm wide for much of its circumference. In *P. equi* the ring consists of a curved band of electron-dense material about 12 nm thick and varying from 55 nm wide on one side of the flagellum to about 120 nm on the other. Some care in the interpretation of the cross-sectional shape and dimensions of the ring is needed, because the angle of the plane of the ring to the long axis of the flagellum can vary from 90° to about 70°. The angle may change during the final stages of differentiation, as reported for *N. patriciarum* (5), or may not have any fixed angle, as reported for *P. mae* and *P. dumbonica* (12). In either case, a degree of flexibility is implied in the perikinetosomal structures that are attached directly or indirectly to the circumflagellar ring.

The component of the harness most closely applied to the kinetosome is the skirt. As first described by Heath et al. (8) for *N. frontalis*, the shape of

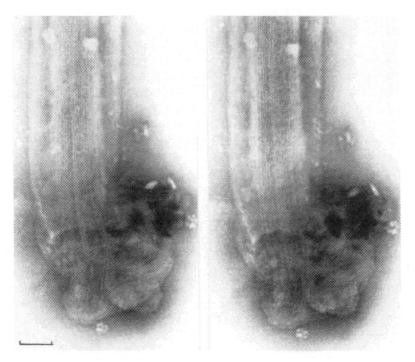

Figure 11   Stereo pair of electron micrographs of a negatively stained kinetosome, some perikinetosomal structures, and proximal end of the flagellum of a zoospore of *P. equi*. The specimen was tilted 12° either side of horizontal in the electron microscope. Scale bar equals 0.1 μm.

the skirt in transverse sections of the kinetosome resembles a somewhat flattened C. This is true for all the isolates so far studied. In *Neocallimastix* the skirt embraces the most proximal end of the kinetosome, to which it is attached in the region of triplets 4–7 (Figs. 5, 10). Two or more struts connect the free arms of the skirt to the circumflagellar ring. The simplest situation seems to occur in *Neocallimastix*, in which there is one strut at the broadened end of the arm adjacent to triplets 2 and 3 and a narrow strut to the paddle-shaped end of the other arm. Three struts have been found in *Caecomyces equi*, *Piromyces mae*, and *P. dumbonica* (12), and in *P. equi* there appear to be four. The struts are not all the same thickness. In *P. mae* and *P. dumbonica* the largest (st1) is at one side of the spur adjacent to triplets 8 and 9. The other two (st2 and st3) are on the other side of the spur. St2 tapers toward the circumflagellar ring; it is fused distally with material of low electron density, the connective (12), which, cloud-like, envelops all the perikinetosomal structures.

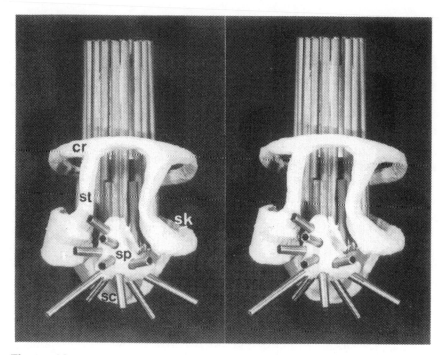

Figure 12 Stereo pair of photographs of a scale model of a kinetosome and peri-kinetosomal harness based on interpretation of electron micrographs of *N. patriciarum*. (From Ref. 5.)

In *P. equi* the four struts are interconnected basally by transverse strips of electron-dense material that in turn are continuous with the skirt and thence are connected to the small scoop underlying the base of the kinetosome. The four struts are on the side of the kinetosome adjacent to the spur. The median pair of these struts (V1 and V2) form a narrow V-shape, with the tip of the V attached to the base of the kinetosome and the tips of the arms ending in the widest part of the circumflagellar ring. The V-struts are roughly triangular in cross-section, with the bases of the triangles facing away from the kinetosome. The two struts (M1 and M2) on either side of the V-pair arise from ribs that extend about 40 nm on either side of the tip of the V and then continue around the base of the kinetosome at a distance of some 30 nm from its surface to join with the skirt. In longitudinal sections this pair of struts are seen to curve away from the kinetosome before joinging with the circumflagellar ring. The other two struts (L1 and L2) arise from the edge of the skirt.

The two arms of the skirt are connected at the base of the kinetosome, which they partly underlie, and are extended proximally to form a cowl-shaped structure, the scoop, which extends about 100 nm below the base of the kinetosome. The opening of the scoop faces the part of the kinetosome that is continuous with doublet 1.

The spur, from which radiate numerous microtubules, lies on the side of the kinetosome faced by the opening of the scoop. In *N. patriciarum* it is roughly triangular (see Fig. 12), with one axis parallel to the long axis of the kinetosome and the base mostly proximal to the kinetosome (Figs. 9 and 10I,J,K). The spur is attached to the proximal edge of the kinetosome by a conical stalk, which also has some connections to the skirt around the edge of the kinetosome. In *P. equi*, the shape of the spur approximates to a slightly hollowed hemisphere, with the somewhat concave base directed toward the kinetosome but at an angle of about 50° to its longitudinal axis. The spur is about 200 nm in diameter centered about 100 nm from the nearest point of the kinetosome. There is no pronounced connection from the spur to the kinetosome, although in most serial sections a tenuous connection can be seen from one edge of the spur to the adjacent part of the harness. In *C. equi* and *P. mae* the shape of the spur is like a longitudinally split cone, with dimensions overall of between 150 and 200 nm. In *P. mae* the spur has a very electron-opaque shell 10–15 nm thick and is joined to the connective by a bundle of fibrillar material.

One of the characteristics of the orders Chytridiales and Spizellomycetales is the presence of kinetosome props (14,15,17). Props are rods, prominent in both longitudinal and transverse sections, extending from a distal extension of each C tubule of the kinetosome triplets to the plasma membrane. Although proplike structures have been seen occasionally in certain planes of sectioning parallel to the longitudinal axis of kinetosomes in *N. patriciarum* (see Fig. 8), these have been shown to be due to fortuitous alignment of parts of the tilted circumflagellar ring (5). No structures compatible with being true props have been seen in sections that pass through the long axes of the kinetosomes in any of the isolates of anaerobic fungi in either free or developing zoospores. Furthermore, with one exception, there are no props described from examination of transverse sections of kinetosomes in zoospores, both immature and free, many as serial sections. The exception is the report by Li et al. (12) for *P. dumbonica* that "nine props are observed in cross-sections of the kinetosome in free zoospores" and that "when present, they are very clear in appropriate planes of section." Surprisingly, this claim was not supported by a figure, and neither of the two whispy strands of material labeled as being props in the one cross section (which is part of a series passing through the base of a flagellum and its kinetosome) that they claimed showed the props connects to a microtubule or, come to that, the plasma membrane. In view of the clarity of props in both longitudinal and transverse sections of zoospores of most of the aerobic chytridiomycetes and the clarity of the components of the perikinetosomal harness in the anaerobic ones, I conclude, on the basis of the present evidence, that props do not exist in the latter group. Their function, presumably skeletal, is apparently met by the interconnections of the circumflagellar ring with the other components of the harness. It is not obvious why such an elaborate structure should have evolved.

Whatever the detailed differences between isolates, the perikinetosomal structures making up the harness conform to a common basic plan and thus constitute an essentially similar entity (5). The point has been made (12) that this implies a monophyletic origin for the anaerobic chytridiomycetes. Appropriate studies of further isolates will determine whether differences in detail, such as the number and form of the struts, are phylogenetically significant.

3. Surface Layers

The layer over the surface of the flagellar membrane is approximately 14–17 nm thick when seen in section, in which it appears electron-dense and closely applied to the membrane (Figs. 4–6). In negatively stained preparations (Fig. 13), this layer is seen to be composed of a regular array of particles. The individual particles, which are about 3.5 nm in diameter, are not seen clearly, but they are arranged in rows to give prominent transverse striations and usually less obvious oblique striations about 77° to the long axis of the flagellum. Very similar arrays occur in the layers coating the flagella of *N. patriciarum, S. communis, P. communis, P. lethargica,* and *P. equi.* The center-to-center spacing of the transverse striations, previously given as 4.3 nm for *N. patriciarum* (4), ranges from 3.3 to 4.8 nm, with most values about 3.9 nm. Bearing in mind the flexibility of the layer needed to accommodate the movements of the flagellum, it is not unexpected that there should be some variation in the spacing. The layer is continuous from the proximal end of each flagellum to the end of the narrow tip and sometimes even continues a little way beyond the termination of the membrane (Fig. 13d). It would appear that some components of the layer that are positively stained by the electron-dense stains (osmium, lead, and uranyl salts) used with sectioned material are readily penetrated by negative stain and therefore not seen by this technique. Similarly, in sections of *P. communis* and *P. lethargica* flagella, additional fibrillar components that extend some 8–12 nm from the general surface of the layer are visible (Fig. 4b,d), but these have not been detected in negatively stained preparations.

When seen on membrane fragments in negatively stained preparations, the cell body surface layer appears to be composed of parallel fibrils approximately 12 nm in diameter and about 17 nm center to center (Fig. 14). From their appearance at higher magnification (Fig. 14c), each fibril could be a

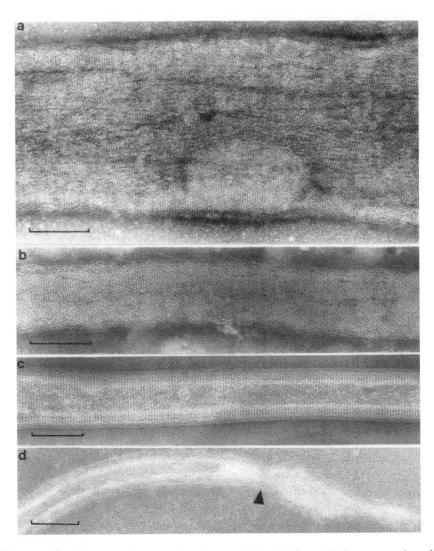

Figure 13   Electron micrographs of parts of flagella from **(a)** *P. communis* and **(b,c,d)** *N. patriciarum*, negatively stained with ammonium molybdate to show the layer coating the surface of the plasma membrane from its proximal end, adjacent to the region shown in **(a)**, to the narrow tip **(b,c,d)**. As shown in **(d)**, the layer sometimes extends beyond the very tip of the membrane (arrowhead). In **(c)** and **(d)** the negative stain is present inside the membrane so that the core appears dark; the membrane is seen as the pair of light bands on either side of the core. Scale bars equal 0.1 μm.

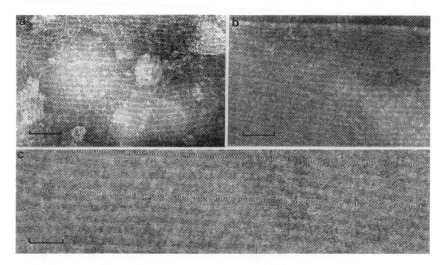

Figure 14 Pieces of plasma membrane from the cell body of **(a)** *N. patriciarum* and **(b,c)** *P. communis* zoospores negatively stained with ammonium molybdate, showing the fibrils composing the surface layer. Scale bars equal 100 nm **(a,b)** and 50 nm **(c)**. ((**a**) and (**b**) from Ref. 6.)

microtubule with a wall about 4 nm thick. This is compatible with their appearance in sections, in which they appear as a row of dots (Fig. 15a,b) or circles (arrowheads, Fig. 15c) when seen in transverse section and give a striated pattern when seen in the plane of the section (Fig. 15c,d,e). Fibrous material, which extends up to 80 nm from the surface of the arrayed fibrils, is visible in some sections.

Preliminary studies indicate differences between *Neocallimastix* species and *Piromyces* species detectable with antisera raised against the surface layers (Table 1), but the level of resolution of ultrastructural studies attempted to date is insufficient to resolve differences between the surface layers from different genera. The application of reconstruction techniques to high-resolution electron micrographs of negatively stained flagella and cell body plasma membrane should reveal details of the structural units of which they are composed.

The two layers, on flagella and cell body, almost meet but do not join at the very base of the flagellum, where in all genera there is a slight narrowing of the flagellum to form a "waist" (Figs. 8, 16). In zoospores of the uniflagellate species the base of the flagellum often lies within an annular groove some 75–90 nm wide (Fig. 16), which in different zoospores ranges in depth

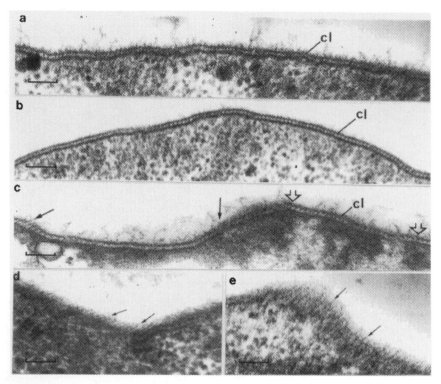

Figure 15 Electron micrographs of sections showing the array of fibrils forming the layer (cl) covering the zoospore cell body. **(a,d)** *S. communis*; **(b,e)** *P. communis*; **(c)** *P. lethargica*. Arrows indicate fibrils in the plane of the section; arrowheads indicate where fibrils appear microtubular. Scale bar equals 0.1 μm. (From Ref. 6.)

Table 1 Responses to Sera Against Zoospore Surface Layers[a]

	N. patriciarum		NRE10		P. communis		P. equi	
	Flag.	Cell	Flag.	Cell	Flag.	Cell	Flag.	Cell
α-N.p.	+ + +	+ + +	+ +	+ +	−	(+)	−	−
α-P.c.	−	(+)	−	(+)	+ + +	+ + +	+ + +	+ + +
α-P.e.	−	−	−	−	+ +	+ +	+ + +	+ + +

[a]Antisera raised separately to glutaraldehyde-fixed zoospores of *Neocallimastix patriciarum* (N.p.), *Piromyces communis* (P.c.), and *P. equi* (P.e.) by the method of Hardham et al. (18) were tested against zoospores of these three species and also zoospores of *Neocallimastix* NRE10 isolated from Svalbard reindeer. The intensity of fluorescence of an FITC-labeled second antibody is shown as: + + +, very strong; + +, medium; (+), very weak and not on all zoospores; −, no fluorescence. Flag., flagella surface; cell, cell body surface. Antibodies in sera to *N. patriciarum* react well with two *Neocallimastix* isolates but negligibly with *Piromyces* spp. Conversely, antibodies to the *Piromyces* species react strongly with each other's surface layers, but very weakly to those of the *Neocallimastix* isolates.
Source: Unpublished data of D. Atkinson, A. Hutchings and E.A. Munn, presented at ISEP Symposium, London 1988.

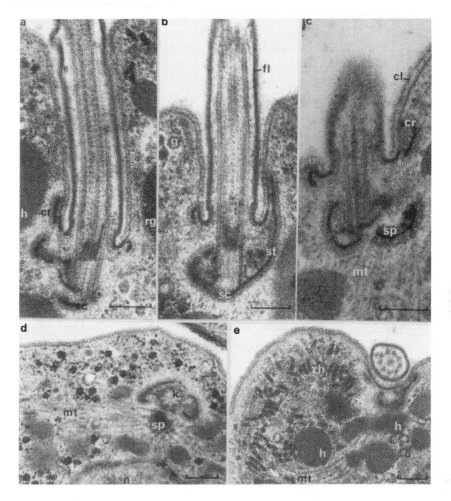

Figure 16 Electron micrographs showing proximal ends of flagella and kineto-somes of monoflagellate zoospores. **(a,b,c)** in longitudinal section; **(d,e)** in transverse section. **(a,b,d,e)** *P. communis*; **(c)** *P. equi*. Scale bars equal 0.2 μm.

from about 0.2 μm to about 1 μm. The extracellular layers coating the zoo-spore cell body and flagellum occupy much of this gap. One lip of the groove is frequently higher than the other and often continues on to form a dome or bulge in the surface of the cell body.

4. *Plasma Membrane Fibrils*

The plasma membrane in or near this domed region of the cell body is char-acterized by the presence of a small array of parallel fibrils, each about 5.5 nm wide and 22 nm center to center, that extend about 40 nm into the cytoplasm. The arrays were first reported for *Neocallimastix* (4) but occur in all genera so far examined (Fig. 17). Typically there are about 20 fibrils in a row, but the overall extent of the array is not known, and more than twice this number has been observed (Fig. 17a). Although these structures look like fibrils, the same appearance would be obtained from end-on views of a sectioned row of parallel ribbons. In either case, they seem to be characteristic components of the zoospores, limited in their position to one small area near the point of insertion of the flagellum. Sometimes one or two rows of electron-dense patches are visible in the cytoplasm immediately adjacent to the plasma membrane fibrils (Fig. 17a,b).

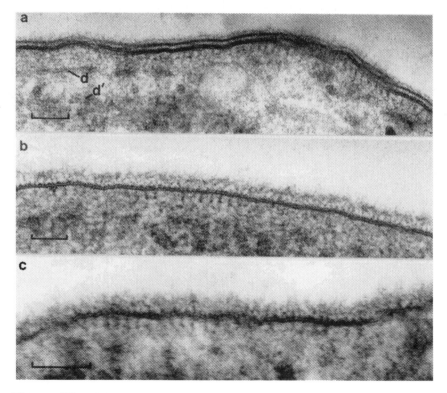

Figure 17 Electron micrographs of sections showing the arrays of fibrils that extend into the cytoplasm from the plasma membrane in a region in the vicinity of the insertion of the flagella. **(a)** *N. patriciarum*; **(b)** *P. communis*; **(c)** *P. equi*; **(d)** *S. communis*. In some sections dense patches (d, d′) occur in a row adjacent to the fibrils. Scale bars equal 100 nm.

5. Microtubule Distribution

In addition to the microtubules in the flagella and kinetosome, zoospores contain arrays of microtubules that radiate into the cytoplasm from the spur, which is evidently a microtubule-organizing center. Although there are minor species differences, in general these microtubules seem to form two groups. One group, the larger, forms a broadly conical or hemispherical array that extends more or less basally (anteriorly). The second group is much smaller and fans out laterally from the kinetosome. This arrangement was first described for *N. frontalis*, in which there are about 14 microtubules directed anteriorly in a cone from every spur and a distinct group of approximately 10 microtubules that form a planar array fanning out close to the plasma membrane from the spurs on four or five kinetosomes at one end of the row. This latter group of microtubules has been designated the posterior fan (8). The precise relationship of the microtubules in the posterior fan with the spurs has not yet been described. In *N. patriciarum*, the longitudinal, distal part of each spur gives rise to microtubules that are mostly directed laterally and the proximal part to microtubules some of which are directed anteriorly and some laterally from the kinetosome (see Fig. 9). In *N. frontalis*, the microtubules in the anteriorly directed group extend for about 2.5 μm, and some are closely associated with a beak-like posterior extension of the nuclear membrane. The nucleus in *N. patriciarum* seems to lack this feature, and any apparent association between microtubules and nucleus in this species is regarded as fortuitous (5). A posterior fan has not been identified with certainty in this species or in *N. hurleyensis*. In the uniflagellate zoospores, however, the posterior fan is clearly present. Thus, in *C. equi*, for example, an array of 8 to 10 microtubules fans out from the distal end of the spur perpendicularly to the kinetosome and extends into an adjacent peripheral region of the cytoplasm (11). Lying parallel to this array, between it and the plasma membrane, are so-called megatubules, each of which is about 46 nm in diameter (measured from a figure in Ref. 11). In *P. mae*, the fan consists of about 20 curved microtubules (see also Fig. 16e) that arise from one side of the distal end of the spur. These also have adjacent megatubules. In *P. equi*, the pattern of microtubule distribution follows that described above, with a small group extending laterally toward the plasma membrane and one group having a conical distribution around the adjacent end of the nucleus.

The degree of interaction between the anteriorly directed microtubules and the nucleus is considered to be of phylogenetic significance, and, as mentioned later, it is one of the characteristics used to distinguish the Spizellomycetales from the Chytridiales (15). In the aerobic members of these orders, the zoospore mitochondria are closely associated with this group of microtubules. In the anaerobic fungi, in which there are no mitochondria, this position is occupied by hydrogenosomes.

6. Hydrogenosomes

Hydrogenosomes were first described, in *N. patriciarum*, as amorphous globules, each consisting of a homogeneous, fairly electron-dense matrix enclosed by a single membrane, mostly associated with the microtubules radiating out into the cytoplasm from each spur. The suggestion that the amorphous globules were hydrogenosomes was made by Heath et al. (8), and this was confirmed, for *N. patriciarum*, by subcellular fractionation and combined enzyme-distribution and electron-microscopic studies (19). Although similar subcellular fractionation studies have not been carried out for other genera of anaerobic fungi, it is widely and reasonably assumed that all morphologically similar structures in the anaerobic fungi are also hydrogenosomes. This assumption is given support by the fact that in all free zoospores studied, these entities occupy the same region of the cell, being mostly associated with microtubules (Figs. 3, 16, 18). The hydrogenosomes are pleomorphic, and it seems likely that in life they do not have a fixed shape. The profile is approximately round or oval and smooth but may have a pointed end or fingerlike prolongation. When present, such prolongations are usually oriented along a microtubule and directed toward the kinetosome (Fig. 18a,b,c). All studies to date support the view that the hydrogenosome in anaerobic fungi has a single peripheral membrane. (It should be noted, however, that a double membrane has been demonstrated for hydrogenosomes in some anaerobic protozoa by means of fixation involving permanganate.) Occasionally the hydrogenosomes appear to have internal membranes (Fig. 18d,e), but most of these can be explained as sections through closely abutting hydrogenosomes or infoldings of the peripheral membrane. The two parts of the infolded membrane may be closely apposed, or they may be curved widely apart to give a profile circular in section and containing part of the cytosol (Fig. 19). It is apparent from serial sectioning that many of the hydrogenosome profiles, which from single sections appear to be separate entities, are in fact joined. The hydrogenosomes may thus form a reticulum. This is seen in extreme form in *N. hurleyensis*, in which the hydrogenosome complex consists of a large, multilobed structure (Fig. 20) that in certain planes of sectioning shows annular profiles with an overall diameter of about 2.5 μm. Perhaps in this isolate the stable form of the reticulum is one in which most of the hydrogenosome components are fused. In *P. mae*, both apparently single, small hydrogenosomes about 0.4 × 0.5 μm across and much larger, irregularly shaped entities (megahydrogenosomes [12]) some 2 μm or more across are present. In *P. dumbonica*, on the other hand, the hydrogenosomes are fairly uniformly tubular.

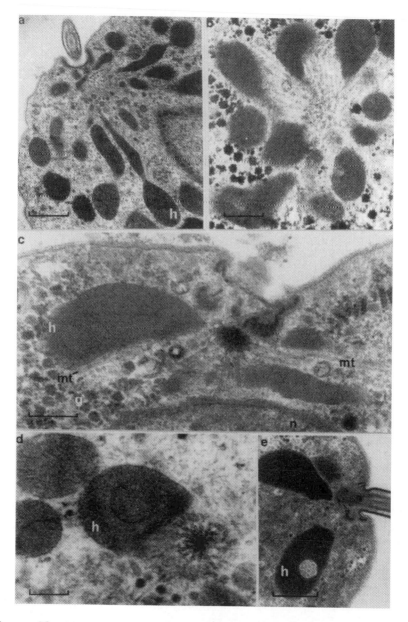

Figure 18 Electron micrographs of sections showing the form and distribution of hydrogenosomes in zoospores of **(a)** *P. lethargica* (NF1); **(b,c)** *P. communis*; **(d)** *P. equi*. The scale bars equal 0.5 μm **(a,e)** and 0.25 μm **(b,c,d)**.

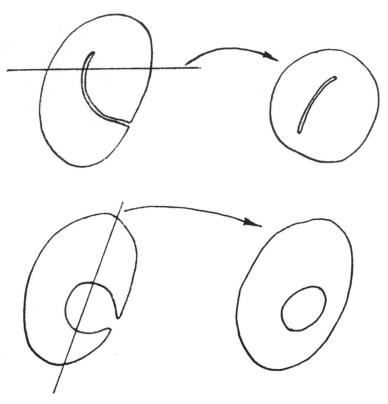

Figure 19 Diagrams to show how sections through infolded surface membrane may give the appearance of internal membrane in a hydrogenosome.

Because of their proximity to the flagella (and by analogy with mitochondria in aerobic zoospores) it may be supposed that the primary function of the hydrogenosomes is to provide energy in the form of adenosine triphosphate (ATP) for driving the flagella. Furthermore, one might expect that the energy required to beat 10 flagella is substantially greater than that needed to drive one flagellum (although it is unlikely to be ten times greater). So one might also expect there to be substantially more hydrogenosomes in a multiflagellate zoospore than in a uniflagellate one. However, assuming equal density of packing, consideration of a simple model (Fig. 21) shows that far more hydrogenosomes can be accommodated *per flagellum* in the cone around one kinetosome than can be accommodated in the corresponding space adjacent to a double row of kinetosomes. It may be calculated, for

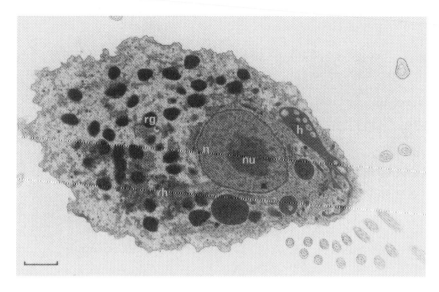

Figure 20 Electron micrograph of section through zoospore of *N. hurleyensis* to show complex form of hydrogenosome (compare Figs. 3 and 22). The scale bar equals 1 μm. (From Ref. 10.)

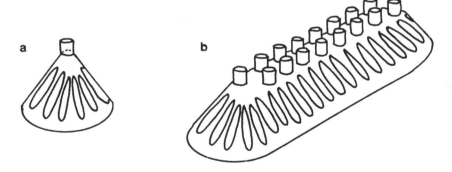

Figure 21 Diagrams to show the space immediately adjacent to **(a)** the kinetosome in a uniflagellate zoospore and **(b)** the double row of kinetosomes in *Neocallimastix*. Assuming hydrogenosomes of the shape depicted, there is room for 14 per kinetosome in the uniflagellates and fewer than 3 per kinetosome in the multiflagellates.

example, that each flagellum in a uniflagellate zoospore (Fig. 21a) will have twice as much volume available to accommodate hydrogenosomes as is available per flagellum in a multiflagellate zoospore having 10 flagella. This ratio increases to 2.5 for a zoospore with 16 flagella (Fig. 21b). Since *Neocallimastix* zoospores are able to swim perfectly adequately, it would appear that potentially the uniflagellates have a generous overcapacity for hydrogenosome-derived ATP production. It would be of interest to calculate the energy demands and compare the total surface area and volume of hydrogenosomes in the two kinds of zoospore (and also to compare these values with data for mitochondria in an aerobic zoospore).

7. Nucleus

The nucleus in the uniflagellate zoospores is typically round or oval, with a generally smooth surface that may be indented or slightly protuberant near the kinetosome. In the zoospores of *Neocallimastix*, on the other hand, the nucleus has substantial prolongations. Thus in the free zoospore of *N. patriciarum,* in which the nucleus is about 2.5–3 μm in diameter and positioned on the flagellar side of the center of the cell (Fig. 3), the nucleus has irregular

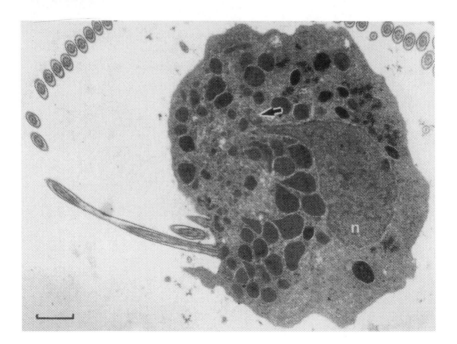

Figure 22 Electron micrograph of section through a zoospore of *N. patriciarum* to show the prolongation of the nucleus (arrowed). The scale bar equals 1 μm.

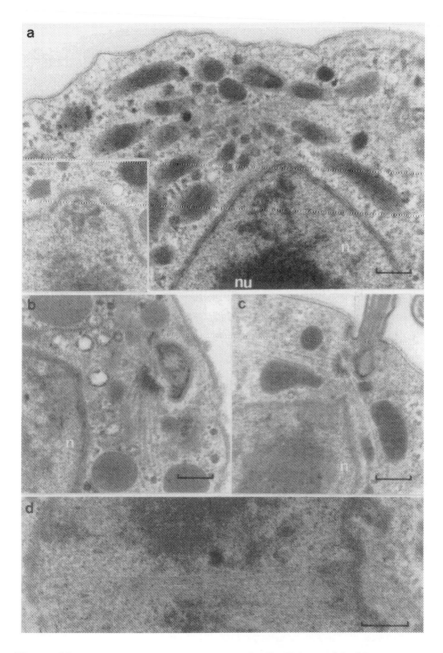

Figure 23 Electron micrographs to show details of the nuclei of the zoospores of **(a,b)** *P. lethargica* (NF1); **(c,d)** *P. equi*. The scale bars equal 0.25 μm **(a,b,d)** and 0.5 μm **(c)**.

prolongations (Fig. 22) that seem to extend in various directions into the cell. In *N. frontalis*, the nucleus has a pronounced conical beaklike extension directed toward the kinetosomes, although it is not known if it consistently points to any one section in the row.

In the uniflagellate zoospores the nucleus has either a rounded protuberance on the side adjacent to the kinetosome, as in *P. lethargica* (Fig. 23a), which often contains electron-dense material (inset, Fig. 23a), or, depending on the plane of sectioning, a blunt point (Fig. 23b) or, as in *P. equi* (Fig. 23c), an indentation with electron-dense material in the adjacent nucleoplasm. In one serially sectioned zoospore of *P. equi*, at least 12 microtubules stretched from the nuclear envelope at the base of the indentation to the opposite side of the nucleus, apparently bisecting the nucleolus (Fig. 23d). Typically, the nuclei contain a large, centrally placed nucleolus, and the peripheral nucleoplasm often contains a ring of eight or so electron-dense patches (see Fig. 24).

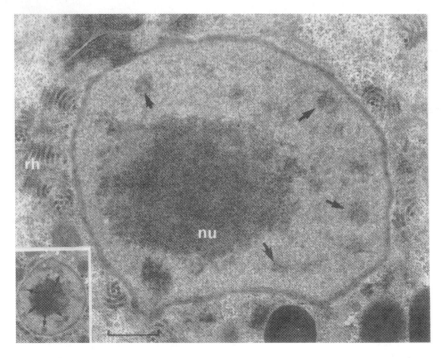

Figure 24 Electron micrographs of sections of nuclei in *N. patriciarum* and (insert) *P. communis*, selected to show the ring of electron densities (arrowed) encircling the large centrally placed nucleolus (nu). The scale bar equals 0.5 μm.

The nucleus in the motile zoospores of *N. patriciarum* (Fig. 24), *N. hurleyensis, Sphaeromonas communis, P. communis, P. equi* (see Fig. 23d), and *P. lethargica* rarely shows a clear limiting membrane, either single or double, but is coated with a fibrous layer. As first reported for *N. patriciarum*, the fibrous layer has a minimum thickness of about 30 nm, with no more than traces of a true limiting membrane visible beneath it. By contrast, the nuclei in sporangia, which in some preparations are present in the same sections as the free zoospores, have both nuclear membranes and pores (Fig. 25).

8. Ribosomelike Particles

The majority of the ribosomelike particles in the zoospores are organized into two structural entities, "helices" and globular aggregates (Figs. 26–28).

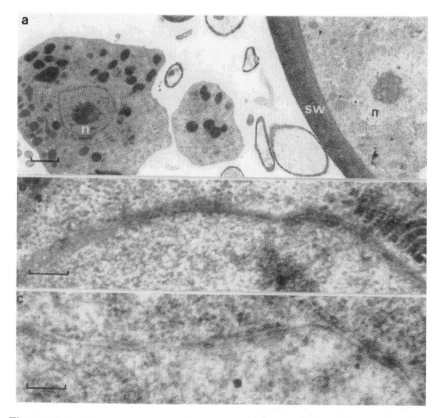

Figure 25 Electron micrographs of **(a)** section passing through a free zoospore and part of a sporagnium of *N. patriciarum*. An area of the surface of the zoospore nucleus is shown enlarged in **(b)** and an area of one of the nuclei in the sporangium in **(c)**. The scale bars equal 1 μm **(a)** and 0.2 μm **(b,c)**. (From Ref. 6.)

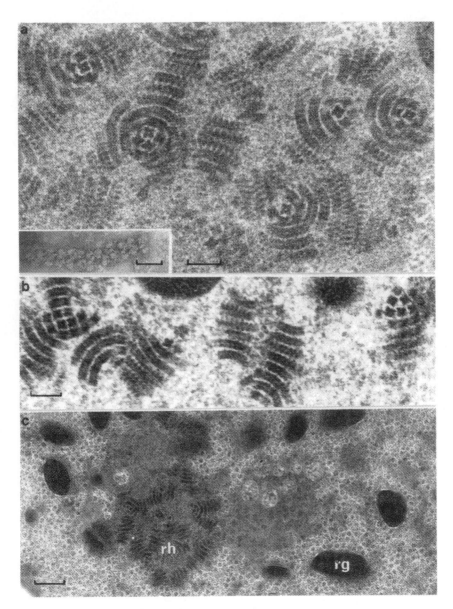

Figure 26 Electron micrographs illustrating the arrays of helices of ribosomelike particles in **(a,c)** *N. patriciarum*; **(b)** *P. lethargica*. The insert in **(a)** is from a preparation negatively stained with 4% ammonium molybdate. The scale bars equal 0.1 μm **(a,b)**, 1 μm **(c)**, and 50 nm (insert).

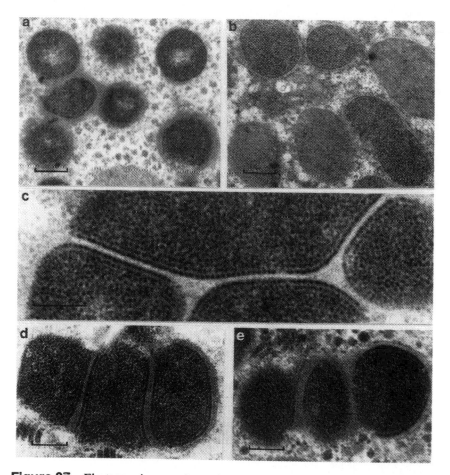

Figure 27 Electron micrographs to show the globular aggregates of ribosomelike particles present in the zoospores of **(a)** *S. communis*; **(b)** *P. lethargica*; **(c,d)** *N. patriciarum*; **(e)** *P. communis*. The scale bars equal 0.2 μm **(a,c,d,e)** and 0.25 μm **(b)**.

The helices commonly occur in arrays in which they present two characteristic profiles, one being of a curved column of 16 nm–diameter particles arranged in pairs at an angle of about 60° to the long axis of the column, and the other the end-on view of this column (Fig. 26). The end-on view is a tetrad of the electron-dense particles in which their centers lie at the corners of a square with sides of 15 nm. When viewed exactly down the axis of the column, a central electron-lucent space is visible at the center of the tetrad (Fig. 26a). Commonly two to four tetrads are grouped together; when four

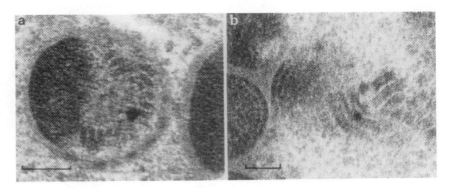

Figure 28   Electron micrographs selected to show continuity between helices and globular aggregates; **(a)** *N. patriciarum*; **(b)** *P. equi*. The scale bars equal 0.2 µm.

are present they form a square array, but the long axes of the columns in such an array are not exactly aligned, and only one of the tetrads in any group reveals the central electron-lucent space (Fig. 26). Columns immediately adjacent to a square array are further out of alignment and are thus seen in oblique section. They in turn are surrounded by helices seen more or less lengthwise. The cytoplasm in the immediate vicinity of these arrays often has an amorphous electron density that distinguishes it from the adjacent cytoplasm (Fig. 26c). Arrays of helices are common near the plasma membrane but also occur in the perinuclear region and adjacent to the globular aggregates.

The globular aggregates of ribosomelike particles are predominantly in the half of the cell away from the hydrogenosomes (Fig. 3), but some are perinuclear, and occasionally they occur adjacent to the hydrogenosomes. Twenty to 30 globular aggregates may be present per section. Each aggregate is approximately ovoid in shape, with maximal dimensions of 0.5 by 1.0 µm (Fig. 27). A number of the sections of aggregates show them associated with or partially bounded by a single membrane enclosed within an amorphous layer also bounded by a single membrane (Fig. 27b).

In virtually all planes of sectioning of the globular aggregates, the particles along the edges are often seen to be in a row (Fig. 27). Occasionally this apparent orderliness is shown by particles within the globules, but the majority do not appear to be arrayed. One explanation for this would be that the particles are in the form of a string and that the string is wound upon itself like a hand-wound ball of knitting wool. This would also account for the fact that the globules all seem to have a defined boundary, although for most of it there is no limiting structure. The formation of such a string implies that each particle has bonding sites on opposite sides with affinity for corresponding sites on two other particles. This is quite different from the bonding pattern involved in the formation of the helices. An alternative model for the linear arrangement in the aggregates would be that the particles are like beads on a thread. Either way, in at least some cases the ribosome helices are continuous with the ribosome globules, examples of this having been found in *Neocallimastix* and *P. equi* (see Fig. 28).

Individual particles, similar in size and appearance to those in the helices and globules, also occur dispersed in the cytoplasm of the zoospores, but the vast bulk of these particles occur in the two organized entities. At the time the zoospore encysts and begins to develop into the vegetative stage, the association between the particles is disrupted and does not reform until late in differentiation of the next round of zoospores (see below). One speculates that while the zoospores are motile, protein synthesis is minimal, and that the helices and globules represent a readily accessible store of ribosomes ready for rapid translation of mRNA and hence production of protein essential for encystment and development of the thallus. In this case the helices and globules would be analogous to the nuclear cap characteristic of Chytridiales. It is not clear why two distinct but clearly associated and sometimes contiguous storage forms of the particles are needed or are beneficial to anaerobic zoospores.

9. Membrane Stacks (and Other Vesicles)

Membrane stacks, first described for *N. patriciarum*, occur in all species so far examined. The stacks consist of four to 15 discoid, flattened membrane vesicles, each some 0.12 to 0.15 μm in diameter (Fig. 29). All the disks in each stack are more or less parallel, in register, and equidistant in the range 20–30 nm center to center. The flattened faces of each disk may be separate, with the space apparently empty (Fig. 29a,b) or filled with electron-opaque material (Fig. 29c), or the faces may be closely apposed (Fig. 29d). The zoospores of *P. mae* and *P. dumbonica* are reported to contain many membrane stacks (12). In *P. mae*, two membrane stacks were found associated with a hydrogenosome, and others were free in the cytoplasm. In *P. equi* and *N. patriciarum* flattened vesicles, resembling those in the membrane stacks, have been seen associated with one edge of crystallites (Figs. 29e, 32a). It is possible that the membrane stacks are the Golgi apparatus, or its precursor, but there is no cytochemical or other evidence to support this contention.

Individual paired membranes are seen occasionally in the cytoplasm (Fig. 29f). (These may well be the foci for longitudinal slits seen in some zoospores, the presence of which may be attributed to inadequate fixation.)

A variety of small vesicles ranging in size from 30 nm to 140 nm, of which some resemble (superficially at least) coated vesicles and some resemble horseshoes, have been noted in sections of zoospores of various species (Fig. 30). Some of the vesicles adjacent to the kinetosomes may be involved in

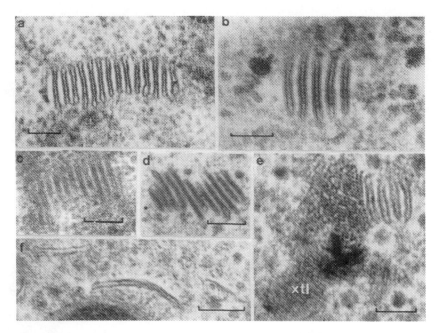

Figure 29   Electron micrographs of **(a–e)** membrane stacks and **(f)** single paired membranes in **(a)** *S. communis*; **(b)** *N. patriciarum*; **(c)** *P. lethargica* (NF1); **(d,e,f)** *P. equi*. In **(e)** the membrane stacks are adjacent to a crystallite (xtl) (see Fig. 32) sectioned gransversely and obliquely. The scale bars equal 0.1 μm **(a–d)** and 0.2 μm **(e,f)**.

membrane synthesis when the zoospores attach (see below), but in general no function has been ascribed to these entities. Oval microbodies some 0.6 by 0.9 μm in diameter with uniformly electron-dense contents and a triple-layered coat occur in *P. dumbonica*. Vesicles resembling autophagous vesicles have also been described (see also Fig. 30).

Groups of entities seen in sections as rings occur commonly in *P. lethargica* (NF1) and *P. equi* (Fig. 31). In the former species the rings sometimes have an ordered arrangement (Fig. 31a,b), but in *P. equi* the rings are scattered in small groups over a considerable area (Fig. 31c). They do not appear to be cross sections of vesicles, but could be very short tubes.

10. Crystallites

Crystallites are more or less ordered arrays of fibrils (Fig. 32), which are sometimes closely associated with flattened vesicles resembling membrane stacks (see Fig. 29e, 32a). They have been seen in sections of the zoospores of *N. patriciarum, S. communis, P. lethargica, P. communis,* and *P. equi.*

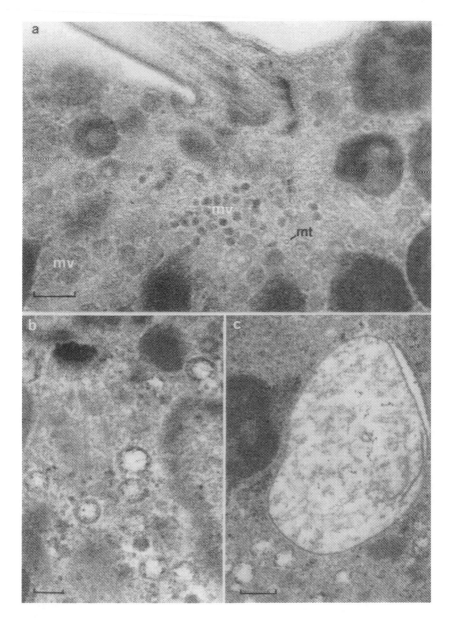

Figure 30 Electron micrographs to show three kinds of vesicle encountered in free zoospores of (a) *N. patriciarum*; (b,c) *P. equi*. The scale bars equal 0.2 μm (a), 0.1 μm (b), and 2 μm (c).

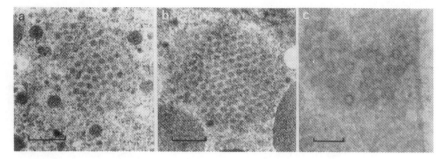

Figure 31 Electron micrographs of rings that occur in large, sometimes organized groups in **(a,b)** *P. lethargica* (NF1) or distributed in scattered small groups in **(c)** *P. equi*. The scale bars equal 0.25 μm **(a,b)** and 0.1 μm **(c)**.

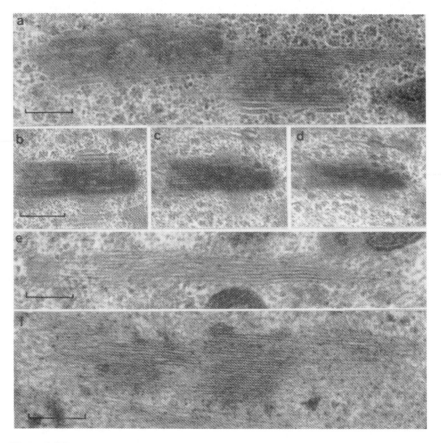

Figure 32 Electron micrographs of crystallites in **(a–d)** *N. patriciarum*; **(e)** *S. communis*; **(f)** *P. equi*. **(b,c,d)** Views of the same crystallite tilted 0°, 36°, and 56°, respectively in the electron microscope. The scale bars equal 0.2 μm.

In *N. patriciarum*, the crystallites are 500–900 nm long and about 150–250 nm wide and show longitudinal and oblique striations (Fig. 32a). The appearance of the striations can be altered by tilting the section in the electron microscope (see Fig. 32b,c,d). In other genera the crystallites consist of bundles of more or less loosely associated fibrils with only partial (e.g., *P. equi*, Fig. 32f) or no obvious (e.g., *S. communis*, Fig. 32e) lateral interaction. These differences may in part be due to the plane of sectioning. It is not known whether the crystallites in different genera are composed of homologous fibrils, but on the basis of appearance, species within a genus have the same kind of fibrils.

11. Glycogen

Zoospores of *N. patriciarum* stain positively for glycogen with Best's carmine, but not if preincubated with diastase (4). Particles with the characteristic appearance of glycogen rosettes are distributed throughout the cytoplasm of the zoospore cell body, occupying all areas not occupied by other structures, and are released from disrupted zoospores examined in the electron microscope by negative staining (Fig. 33). It is reasonable to presume that the function of the glycogen, laid down during vegetative growth, is to serve as an energy resource for the zoospore while it is motile and during

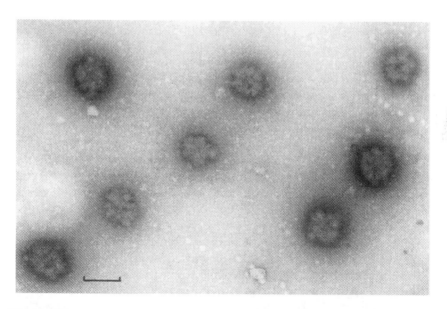

Figure 33 Electron micrograph of glycogen rosettes from *N. patriciarum*, negatively stained with 2% sodium phosphotungstate. The scale bar equals 0.1 µm. (From Ref. 4.)

encystment. In this role it would fulfill the function of the lipid, stored in globules, that characterizes the zoospores of some aerobic fungi.

B. Encysting Zoospores

From light microscopy it is known that the first stages in encystment are attachment to a suitable substrate, rounding up, and, in most species, loss of flagella. Ideally, studies of the sequence of events in encystment would use synchronized cultures of zoospores. In the absence of this ideal it has been possible to determine the progress of the ultrastructural changes by studying zoospores of *N. patriciarum*, fixed at intervals after being brought into contact with a surface, and noting the most advanced change at each time interval (E. A. Munn, C. G. Orpin, and C. A. Greenwood, unpublished observations). Changes are seen in the nuclear envelope, in the distribution of the ribosomelike particles and hydrogenosomes, and in the zoospore cell body surface layers. These are followed by development of the rhizoid, which marks the beginning of vegetative growth.

1. Loss of Flagella

The most remarkable thing about the loss of the flagella from encysting zoospores, which has been studied in particular in *N. frontalis* (20), is that there is also loss of the kinetosomes. This might be suspected from examination of material in the light microscope using phase contrast, in which a knob can be seen at the basal end of each free flagellum when these are detached by shearing — for example, by rapid passage through a narrow pipette tip. It has been confirmed with the electron microscope by examination of zoospores of *N. frontalis* that had encysted within a sporangium and thus provided many samples in which the detached flagella with kinetosomes remained adjacent to the zoospores (20). Each kinetosome carried with it at least the skirt and spur components of the harness but not, apparently, any of the microtubules arising from the spur or, surprisingly, any membrane. The implications of this are twofold. First, as emphasized by Heath et al. (20), it implies that the kinetosomes are not autonomous but arise de novo during zoosporogenesis. This fits in with the observed absence of centrioles at the mitotic spindle poles in sporangia (21) (see below). Second, to avoid formation of a very substantial hole, or in the case of *Neocallimastix* up to 20 very substantial holes, it implies that a mechanism must exist for rapid fusion of the plasma membrane around the site of each deleted kinetosome. It is possible to envisage processes whereby the kinetosome is detached and the plasma membrane seals to retain the integrity of the cell, and one is shown diagrammatically in Fig. 34. The envisaged process fails to account for the absence of membrane around the kinetosome and thus is at odds with the data of Heath et al. (20). As a compromise to try to meet this difficulty, the membrane

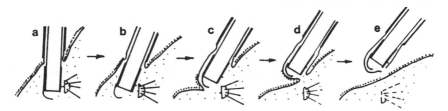

Figure 34 Diagram to illustrate a possible mechanism for sealing the plasma membrane of encysting zoospores, needed when both the flagella and kinetosome are lost. **(a)** The situation in the free zoospore; **(b)** at the first stage the spur becomes detached from the kinetosome, which begins to tilt and move away from the spur; **(c)** movement of the kinetosome continues and the plasma membrane begins to fold in under the base of the kinetosome; **(d)** this process continues, and possibly at this time the microtubules, and perhaps the spur, begin to dissociate. **(e)** The process of detachment is completed with fusion of the plasma membrane.

around the kinetosome is depicted as not sealed (Fig. 34e). The new plasma membrane could be derived from the numerous vesicles present at the base of the kinetosome (see Fig. 30), but there is no evidence to support this conjecture. It is difficult to imagine what overall benefit the organism could derive from shedding flagella and kinetosomes.

2. Surface Layers

Changes in the surface layers of the attached, rounded zoospore cell body are shown in Fig. 35. Essentially they consist of development of a cell wall and development of numerous fingerlike prolongations of the plasma membrane coated with the surface layer. The nascent cell wall is first detectable as a thin layer between the plasma membrane and its surface layer. It is further thickened, apparently, by deposition of layers on the inner surface, a process that continues throughout the period of vegetative growth until the contents of the sporangium become separated from the rhizoidal system (see below).

3. Ribosomelike Particles

From the earliest stages of encystment there is a reduction in the number of helical arrays of ribosomelike particles, and with time these completely disperse. Then there is dispersal of the globular aggregates of ribonucleoprotein particles and appearance of rough endoplasmic reticulum (Fig. 36).

4. Nuclear Envelope

Changes in the nuclear envelope are among the earliest detectable in the encystment process. The fibrous layer coating the nuclear envelope disappears,

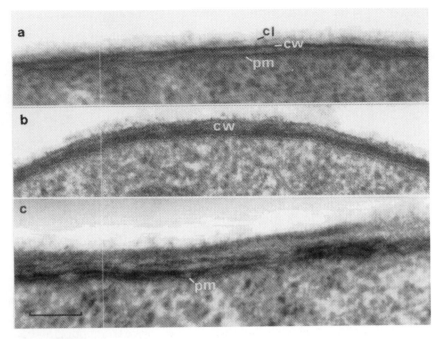

Figure 35 Electron micrographs showing the early stages of (a) formation and (b,c) thickening of the cell wall (cw) between the plasma membrane (pm) and cell surface layer (cl) of encysting zoospores of *N. patriciarum* obtained from samples fixed at intervals after bringing them into contact with an agar surface under conditions favoring encystment. The scale bar (for all three images) equals approximately 0.1 µm.

and a double nuclear membrane with numerous nuclear pores becomes prominent (Fig. 36c).

C. Vegetative Growth

The overall form of the rhizoids, apparent from light microscopy, is either filamentous and branched, as in *Neocallimastix* and *Piromyces*, or spherical, as in *Sphaeromonas* (23-26) and *Caecomyces* (11). This is determined by the way in which the cell wall is laid down. Although electron microscopy of thin sections can provide information on the structure of the cell wall, including that at the growing tips of the rhizoids, it has not yet been used in any investigation of the factors determining how the cell wall components are assembled. The study of the vegetative stages is one area in which the technique of thin sectioning has been applied to the study of natural popu-

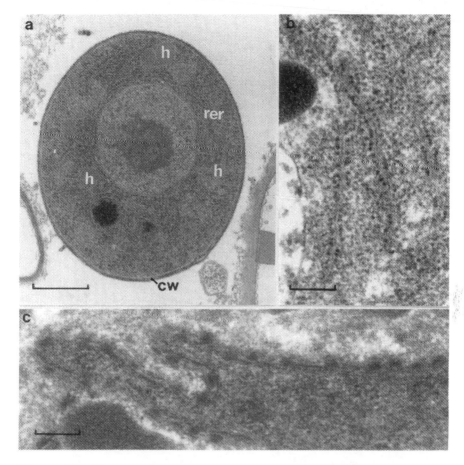

Figure 36 Electron micrographs showing **(a)** an encysted zoospore of *P. communis*. The cell wall (cw) is forming, the hydrogenosomes (h) are dispersed, and large amounts of rough endoplasmic reticulum (rer) are present. **(b,c)** Electron micrographs of encysted zoospores of *N. patriciarum* showing rough endoplasmic reticulum and nuclear membranes, respectively. The scale bars equal 0.5 μm **(a)**, 2 μm **(b)**, and 0.2 μm **(c)**.

lations. The limited information available indicates that observations made on material cultured in vitro are valid.

1. Rhizoids

Little work has yet been done on the formation of rhizoids, and the earliest detected stages that have been examined by electron microscopy, of encysted zoospores of *N. patriciarum*, do not tell us a great deal except that longitu-

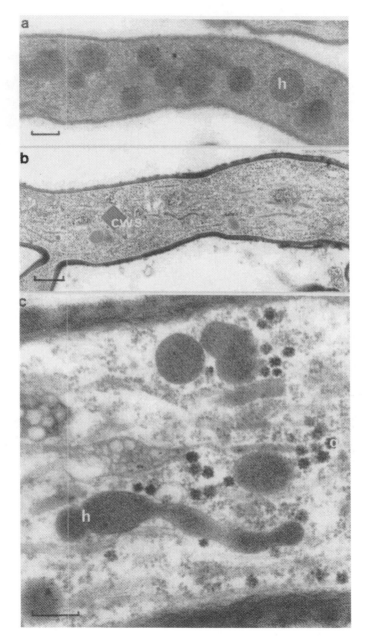

Figure 37 Electron micrographs of longitudinal sections through filamentous rhizoids of **(a)** *P. equi*; **(b,c)** *P. communis*. The scale bars equal 0.25 μm **(a)**, 2 μm **(b)**, and 0.2 μm **(c)**.

dinally arrayed microtubules are present and that the cell wall near the tip of the rhizoid appears very thin. Many more samples are needed, both to determine whether these observations are representative and to study the earlier stages in the formation of the rhizoids.

Longitudinally arrayed microtubules are abundant in the core region of the developed rhizoids or thalli of the genera in which these are filamentous (Fig. 37). Frequently associated with the microtubules, but also apparently lying free in the cytoplasm, are organelles interpreted as being hydrogenosomes. These seem to fall into two main groups. One group consists of electron-dense, more or less tubular forms that tend to lie along the long axis of the filamentous rhizoids. The other group consists of less electron-dense, larger-diameter organelles with oval or circular profiles (Figs. 37, 38). The latter group most closely resemble the organelles identified as hydrogenosomes in zoospores. Both kinds are distributed throughout the length of filamentous rhizoids and around the rim of the spherical rhizoids of *Sphaeromonas*. The bulk of the space within the latter is occupied by widely dispersed wispy material (Fig. 38a). Many of the tubular rhizoids of the other monocentric genera contain small vacuoles and vesicles and arrays of membranes (Fig. 39). Some rhizoids in older cultures are likely to be undergoing degenerative changes subsequent to the separation of the sporangium (see below). These may well be ones that contain many vesicles and in which the contents

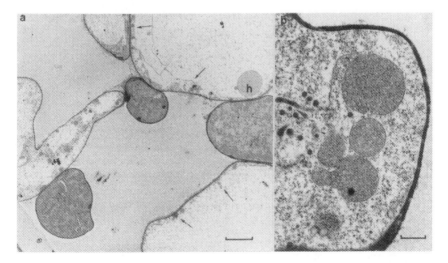

Figure 38   Electron micrographs of sections of the spherical rhizoids of *S. communis*. The scale bars equal 5 μm **(a)** and 1 μm **(b)**. **((b)** from Ref. 6.)

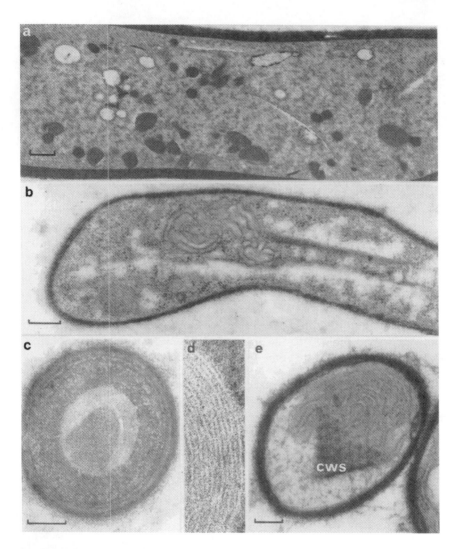

Figure 39 Electron micrographs to illustrate the range of membranous structures, vacuoles, and vesicles encountered in the filamentous rhizoids of *N. patriciarum*. From its overall diameter (about 0.9 μm), **(c)** is assumed to be very near the tip of a rhizoid. The scale bars equal 2 μm **(a)**, 0.5 μm **(b)**, and 0.2 μm **(c,e)**. The magnification of **(d)** is about four times that of **(c)**.

are generally lacking in electron density, but in the absence of systematic studies nothing definite can be said about this. What can be stated with assurance, however, is that all the rhizoids contain characteristic entities called crystals with spots (8) or cubic crystal bodies (12).

2. "Crystals with Spots" and Crystallites

Entities resembling zoospore crystallites occasionally occur in sporangia and rhizoids (Fig. 39a), but as may be seen from the electron micrographs in Figs. 40 and 41, "crystals with spots" are distinct entities. They have been reported as occurring in the vegetative stages of all anaerobic fungi so far examined. Most commonly seen in the rhizoids, they also occur in newly formed sporangia.

Figure 40 Electron micrographs of crystals with spots in rhizoids of **(a)** *N. patriciarum*; **(b,e)** *P. communis*; **(c)** *S. communis*; and **(d)** *P. equi*. The scale bars equal 0.2 μm. (**(a,b,c)** from Ref. 6.)

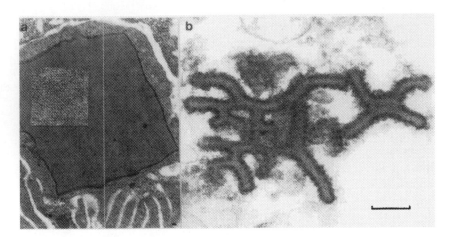

Figure 41 Electron micrographs of sections of crystals with spots from *N. patriciarum*. The crystal in (**a**) is 3.6 μm long. The inset shows an enlargement of a portion of the crystal pattern, correctly oriented, at the same magnification as that in (**b**). The scale bar in (**b**) equals 0.25 μm.

The overall dimensions of crystals with spots usually range from about 0.3 μm to about 0.8 μm, although the largest one so far encountered, in *N. patriciarum*, was 3 × 3.6 μm (Fig. 41a). From the variety of profiles seen in sections, with at least two opposite edges parallel, it appears that the basic shape of the crystals is that of a parallelepiped. This regular form is distorted, however, particularly at the corners, by the presence of finger-like prolongations. The prolongations are sometimes seen to be branched, and an extreme, almost fractal-like development of this is shown in Fig. 41b.

Irrespective of the shape of the edges, the surface of the crystals is coated with a regular array of particles ("spots") each some 20–30 nm in diameter. When lying in the plane of the section this array is seen to be hexagonal, with the particles about 39–45 nm center to center. The shape of the particles is not clear; they are roughly hemispherical seen edge-on and sometimes appear to be six-pointed stars when seen *en face*. In early sporangia of *P. mae* (12), the particles are seen to be regularly interconnected by fine fibrils (see also Fig. 40e). The crystals look as though they might be coated by a membrane, but I think that the particle array is applied directly to the surface of the crystal. The apparent repeat of the crystal lattice is 6.5–10.4 nm, depending on the plane of sectioning.

D. Zoosporogenesis

For convenience of description of the ultrastructural changes, the differentiation of zoospores within sporangia has been divided into seven sequential stages (Fig. 1). The sequence is based on logical interpretation of electron micrographs of sections of samples containing many stages of development of the sporangia, correlated with observations on living material. The sequence is based primarily on studies of *N. patriciarum*. More limited data from similar studies of *P. communis, S. communis, P. lethargica,* and *P. equi* are fully compatible with the observations on the multiflagellate species, as are data from studies of individual sporangia of *N. frontalis* selected from live cultures and then fixed (8).

1. Stage 1

The first sign of the onset of zoosporogenesis is nuclear division within the enlarged, or enlarging, sporangium. Studies of nuclear division in the anaerobic fungi are limited to monocentric members of the group, and the only study done in any detail is that performed by Heath (21,22) on *N. frontalis,* supplemented with observations on another isolate, *Neocallimastix* PN2. In these, mitosis and spindle formation are entirely intranuclear, and overall the mechanism closely resembles that reported for other fungi in which centrioles persist throughout nonmotile stages of the life cycle (27,28). The limited information on these processes in *N. patriciarum* and uniflagellate members of the group indicates that this is true of them also.

For *N. frontalis,* Heath and Bauchop (22) distinguished between early (young) and late (mature) developing sporangia, depending on their assessment, based on abundance of microbodies in the cytoplasm and thickness of the sporangium wall, of how near the organism was to zoosporogenesis. In early development, the interphase nuclei are distinguished by the presence of a single osmiophilic disk in a depression in the cytoplasm at one end of the nucleus. This nucleus-associated organelle (NAO) (Fig. 42a) is replaced in late development by a cluster of centrioles (which eventually become the zoospore kinetosomes). The nucleoplasm adjacent to the NAO is differentiated into a hemispherical zone of low electron density. At the interface of this zone with the rest of the nucleoplasm there are approximately 20 small electron-dense globules that are interpreted as being kinetochores with attached chromatin. Occasionally, short lengths of microtubules (kinetochore microtubules) are seen terminating at the kinetochores. The onset of mitosis is marked by duplication and separation of the NAOs (Fig. 42b,c) and the associated regions of the nuclear envelope. Concomitantly, the adjacent region of nucleoplasm elongates, and nonkinetochore microtubules appear within it. As the NAOs separate further, a bipolar spindle consisting of a tightly packed array of nonkinetochore microtubules interspersed with kinetochore microtubules of various lengths develops (Fig. 42d). The kinetochores apparently double in number. Spindles ranging in length from 0.8 to 3.1 μm are interpreted as metaphase spindles. Chromosomal heterochromatin is

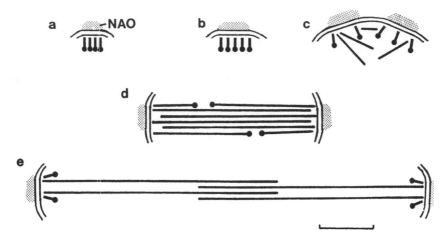

Figure 42   Diagram to illustrate the formation and behavior of kinetochores and spindle microtubules during nuclear division in *N. frontalis*. In late stages of development the osmiophilic nucleus-associated organelle (NAO) is replaced by a cluster of centrioles. The scale bar equals 1 μm. (Reproduced, with slight modification, from Ref. 22.)

spread over much of the length of the metaphase spindle, although other regions of the nucleus also contain apparent heterochromatin. At telophase the spindle is about 7.8 μm long (Fig. 42e); there are clusters of approximately 20 kinetochores at each spindle pole and two half-spindles, each of approximately 28 nonkinetochore microtubules, that interdigitate in the isthmus at the center of the nucleus. Karyokinesis is apparently accomplished by simple median constriction without expulsion of an interzone region. The nucleolus, which persists throughout mitosis, also undergoes medial constriction at karyokinesis. Finally, an aggregate of closely packed osmiophilic globules of characteristic appearance but unknown function persists throughout mitosis and divides with the nucleus at telophase.

The end of stage 1 is marked by formation of a cross-wall separating the sporangium contents from the rhizoid. This is probably concomitant with, or immediately preceded by, the end of nuclear division. The laying down of the cross-wall is presumed to be a rapid process, since although in some instances (Fig. 43) parts of it are very thin, no examples have been found of an incomplete one, which would indicate that it had been fixed during formation. Further, it is presumed that formation of the cross-wall is associated with infolding of the plasma membrane, which must meet and fuse to enclose the sporangium contents.

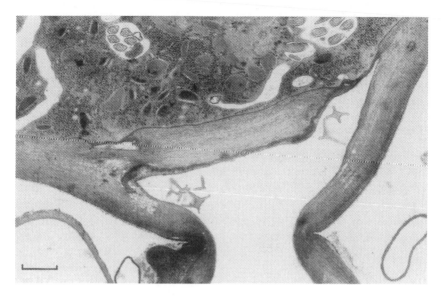

Figure 43   Electron micrograph of a section showing the cross-wall separating the developing sporangium from the rhizoidal system. This sporangium is at stage 4 of zoosporogenesis (see text). Note the crystals with spots present in the otherwise empty rhizoid adjacent to the cross-wall. The scale bar equals 1 μm.

During stage 1 the nuclei occur throughout the sporangium, with some tendency to concentration toward the periphery. The nuclear membranes and nuclear pores are prominent. Ribosomes, either free or membrane-attached, and hydrogenosomes are dispersed throughout the cytoplasm. Microtubules occur, but no pattern in their distribution or any particular association with other structures has been discerned. Crystals with spots are occasionally present.

2. Stage 2

The presumptive cleavage vesicles are first detected at the ends of the kinetosomes away from the nuclei. The surface of each vesicle adjacent to the kinetosomes is flat except where the kinetosomes themselves cause small bumps. The remainder of the vesicle has a convoluted surface. The very small space within the vesicles is mostly filled with fingerlike prolongations of the convoluted surface and small particles.

Some ribosomes, apparently in the form of polysomes, are concentrated adjacent to the nuclei, but most, as in stage 1, are distributed through the cytoplasm.

3. Stage 3

In stage 3 the cleavage vesicles elongate in a direction away from the nuclei. The distance between the kinetosomes and the nuclei increases. The flagella begin to form, extending into the elongating cleavage vesicles, but the flagellar plasma membrane does not at this time show the surface layer characteristic of the mature zoospore.

4. Stage 4

In stage 4 the flagellar surface layer becomes apparent, and the flagella and the cleavage vesicles continue to elongate. The first signs can be seen of polymerization of the cell body surface layer in patches on the cleavage vesicle membrane, which is thus to become plasma membrane (Fig. 44). The distance between the nuclei and their kinetosomes continues to increase. The perikinetosomal structures making up the harness begin to be formed. Although

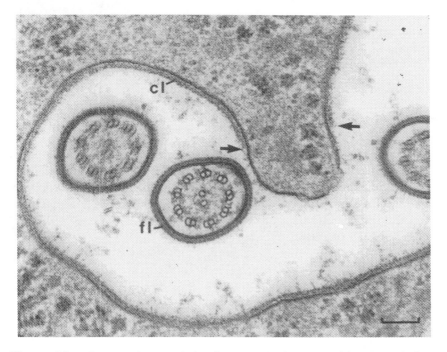

Figure 44 Electron micrograph showing part of a cleavage vesicle at stage 4 of zoosporogenesis. The layer (fl) coating the surface of the flagella is fully formed. The layer (cl) coating the surface of the vesicle, which will become the surface of the zoospore cell body, is only partially formed (arrows indicate the ends of the polymerized regions). The scale bar equals 0.1 μm.

most of the ribosomes are still evenly dispersed throughout the cytoplasm, some globules and a few helices begin to form.

5. Stage 5

In stage 5 the cleavage vesicles have elongated to such an extent that their membranes, now virtually completely coated with the surface layer, are able to fuse with the peripheral plasma membrane (Fig. 45).

6. Stage 6

In stage 6 most of the peripheral plasma membrane is separated from the sporangium wall. Deposition of the cell body surface layer begins on this portion of the plasma membrane. The hydrogenosomes begin to be concentrated between the nuclei and the kinetosomes, and the ribosome globules begin to be concentrated toward the other pole of the nuclei.

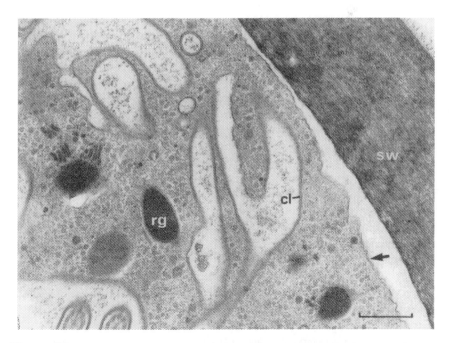

Figure 45   Electron micrograph of a section of a sporangium of *N. patriciarum* showing close proximity of cleavage vesicle membrane, which is coated with surface layer (cl), with the peripheral plasma membrane (arrowed), which is detached from the sporangium wall but not yet coated with the surface layer. The scale bar equals 0.5 μm.

7. Stage 7

In stage 7 the plasma membrane becomes completely detached from the
sporangium wall and becomes fully coated with its surface layer. The zoospores become rounded. The hydrogenosomes become concentrated at the
flagellar pole. The last of the ribosome helices form, and they and the ribosome globules are concentrated around the nucleus away from the kinetosomes. (It would be extremely interesting to know the mechanism by which
these organelles become segregated.)

Finally, the sporangium wall degenerates to leave only the internal and
external layers, and these rupture to release the mature zoospores.

III. TAXONOMY

Differences in ultrastructure of zoospores have been used to divide the Chytridiomycetes into two orders, the Spizellomycetales and the revised Chytridiales (15). Similarities in ultrastructure of the zoospores have been used as
a basis on which to include the anaerobic zoosporic fungi in the Spizellomycetales, the diagnosis of which had to be emended to accommodate multiflagellate forms (8). The characteristics of these two orders as originally
described are set out in the first two columns of Table 2. The description of
the Spizellomycetales was emended by Heath et al. (8) as follows: Nucleus
of the zoospore associated directly or indirectly with the kinetosomes; microtubules radiate into the zoospore from near the kinetosomes; ribosomes
dispersed in the cytoplasm, either singly or in many small membrane-bound
aggregates; mitochondria and microbody–lipid globule complex absent, or
if present, not intimately associated; rumposome absent (column 3, Table 2).
Presumably the change of the word "kinetosome" to the plural "kinetosomes"
was made because the emendment was based on the study of *Neocallimastix*.
The two significant changes were the description of the ribosomes as being
dispersed either singly or in many small membrane-bound aggregates and
the possible absence of mitochondria and microbody–lipid globule complex.

Although, by contrast to the situation in the Chytridiales, the ribosomes
(strictly, the ribosomelike particles) are dispersed in the cytoplasm, they do
not occur as membrane-bound aggregates. As described above, in the free
zoospore they occur predominantly and *characteristically* as groups of columns of tetrads (see Fig. 26) and as globular aggregates that may be associated at one end with a membrane and an amorphous layer (see Fig. 27).
The absence of mitochondria seems such a fundamental difference that to
offer it only as a possible alternative is a failure to recognize its importance;
particularly when it is coupled with the fact — proposed by Heath et al. (8)
and established for *Neocallimastix patriciarum* (19) — that these fungi contain hydrogenosomes.

Table 2 Classification of Chytridiomycetes on the Basis of Zoospore Ultrastructure

Chytridiales (emended, Barr [15])	Spizellomycetales (Barr [15])	Spizellomycetales (emended Heath et al. [8])	Anaeromycetales (proposed ord. nov.)
1. Ribosomes packaged in core of zoospore by a double membrane system.	1. Ribosomes dispersed in the cytoplasm.	1. Ribosomes dispersed in cytoplasm, either singly or in many small membrane-bound aggregates.	1. Ribosomes dispersed in cytoplasm in arrays of columns of tetrads and in globular aggregates associated at one edge with a membrane and amorphous layer.
2. Typically a single, laterally placed microbody–lipid globule complex.	2. Usually a variable number of lipid globules in the anterior, or an aggregate of many small globules.	2. Microbody–lipid globule complex absent or present.	2. Lipid globules absent.
3. Rumposome present.	3. Rumposome absent.	3. Rumposome absent.	3. Rumposome absent.
4. Microtubules originate from one side of kinetosome, run parallel to each other, and extend to edge of rumposome.	4. Microtubules radiate in various directions into zoospore body from near the kinetosome.	4. Microtubules radiate into the zoospore from near the kinetosome.	4. Microtubules radiate into the zoospore from a focus associated with kinetosome.
5. Mitochondria not intimately associated with microbody–lipid globule complex.	5. Mitochondria are associated with the microbody–lipid globule complex.	5. Mitochondria present or absent; if present, not intimately associated with microbody–lipid globule complex.	5. Mitochondria absent.
6. Nucleus not associated with kinetosome.	6. Nucleus morphologically associated with kinetosome, either adjacent to it or connected by microtubules.	6. Nucleus associated directly or indirectly with the kinetosomes.	6. Nucleus loosely associated with kinetosome.

Table 2 (Continued)

Chytridiales (emended, Barr [15])	Spizellomycetales (Barr [15])	Spizellomycetales (emended Heath et al. [8])	Anaeromycetales (proposed ord. nov.)
7. Nonfunctional centriole lies parallel to kinetosome and a fibrous bridge connects them along full length of adjacent side.	7. The nonfunctional centriole is often at a distinct angle to the kinetosome and not connected all along adjacent edges.		7. No nonfunctional centriole.
			8. Hydrogenosomes present; usually associated with microtubules.
			9. Kinetosomes encased in harnesslike structure consisting of skirt, scoop, struts, and circumflagellar ring and attached spur.
			10. Distinct layers of arrayed particles coat surface of plasma membrane over flagella and cell body; layers almost meet but do not join at base of flagella.
			11. May be multiflagellate.
			12. Glycogen particles (rosettes) present.

Other ultrastructural differences between the anaerobic and aerobic chytridiomycetes that appear phylogenetically and taxonomically significant are the presence of the harnesslike structure, consisting of scoop, skirt, struts, and circumflagellar ring, encasing each kinetosome; the absence of nonfunctional centrioles; the absence of centrioles during mitosis (22); the presence of the characteristic layers on the outer surface of the plasma membrane covering the flagella and the cell body of the zoospore (see Figs. 13–16); the presence of the patch of fibrils on the inner surface of the plasma membrane (Fig. 17); and the presence of glycogen. To this list may be added the occurrence in the thalli and early sporangia of the entities described as crystals with spots (see Figs. 40 and 41). Furthermore, leaving aside discussion of cause and effect, all these ultrastructural differences are associated with the restriction of this obligate anaerobic group to the digestive tract of herbivorous mammals. Two of these properties, occurrence as obligate anaerobes in the digestive tract (of vertebrates), and plasma membrane coated with elaborately organized fibrous material, have been given as characteristics in defining a family, the Neocallimasticaceae, within the emended Spizellomycetes and a genus, *Neocallimastix*, respectively (8). Both these usages seem somewhat restrictive, since these properties are characteristic of all the anaerobic chytridiomycetes. Taking all these factors into account, it is proposed that a new order, for which the name Anaeromycetales is suggested, be erected to accommodate this group, as shown in Table 2. As part of this proposal, the Spizellomycetales would revert to the original description (15), and the emendment proposed by Heath et al. (8) would be abandoned. (The name that I would have preferred to suggest for the new order was Enteromycetales, to reflect the fact that all the members dwell in the gut. However, the genus name *Enteromyces* [29] already exists for a member of the Eccrinales. The generic name *Anaeromyces* has been assigned to a polycentric anaerobic fungus from the rumen [30].)

It seems to me that the possession of multiflagellate or uniflagellate zoospores is not a trivial difference, and that these constitute distinct groups of anerobic fungi. Although specimens of the uniflagellate species are found occasionally with more than one flagellum, this is not a general rule. Those (very few) specimens that have been examined by electron microscopy do not show evidence of order in the kinetosomes when more than one is present. This is in contrast to the ordered double-row array of the kinetosomes in *Neocallimastix* species. The view that the uniflagellates and the multiflagellates are distinct groups is supported by the limited data from serology (Table 1). As noted above, a family Neocallimasticaceae has already been erected (8), and this, with emendment, could be retained to accommodate the species of *Neocallimastix*. A second family could then be erected to accommodate the uniflagellate species.

The ultrastructural uniqueness of the anaerobic chytridiomycetes, their complete adaptation to the digestive tract of herbivorous mammals (which includes their requirement for temperatures of 37–39°C), and their distribution throughout a phylogenetically diverse range of groups (perissodactyl and artiodactyl ungulates, elephants, and Australian marsupials have to date been shown to be hosts) implies that they could have existed as a separate group since the time that these mammals began to diverge, at least 120 million years (some 4×10^{10} generations of fungi) ago. Considering the diversity and geographical distribution of their hosts, the comparative uniformity of ultrastructure of the anaerobic chytridiomycetes is remarkable. While it might be argued that the environment within the hosts is relatively uniform, this itself cannot account for uniformity in ultrastructure, since a range of genera with distinctive ultrastructural features occur within the same host.

IV. CONCLUSION

ⵏhe study of the ultrastructure of the anaerobic fungi has contributed to their characterization since the discovery of the first members of the group in the rumen (31), and indeed, it has contributed, in a minor way, to their recognition as fungi. The similarity in detail of a wide range of characteristic ultrastructural features of the anaerobic fungi from a range of herbivorous mammals indicates that they constitute a distinct phylogenetic group. Some of these features, such as the possession of hydrogenosomes and the absence of mitochondria, can be correlated with their unique physiological and metabolic properties. It is hoped that further ultrastructural studies will be undertaken to seek further correlations between structure and function. I have indicated a number of areas in which careful study is needed. While it is likely that the techniques of thin sectioning and negative staining, whose value has been demonstrated in this chapter, will continue to yield valuable data, it is anticipated that the additional application of the techniques of freeze-fracture, deep etching, and immuno-electron microscopy, coupled where possible with subcellular fractionation studies and the newer high-resolution light microscopy techniques, will dramatically advance our understanding of this fascinating group.

ABBREVIATIONS USED IN FIGURES

c, cell wall
cl, layer coating surface of zoospore cell plasma membrane
cr, circumflagellar ring
cw, cell wall

cws, crystals with spots
cy, cylinder/spiral
d_1-d_9, outer doublets of axoneme
fl, layer coating surface of flagellar plasma membrane
g, glycogen
h, hydrogenosome
k, kinetosome
ms, membrane stack
mt, microtubules
mv, microvesicle
n, nucleus
nu, nucleolus
p, plug at base of cylinder/spiral
p', basal prolongation of plug
pmf, plasma membrane fibril array
rg, globular aggregates of ribosomelike particles
rh, helical arrays of ribosomelike particles
sc, scoop
sk, skirt
sp, spur
st, strut
sw, sporangium wall
v, vesicle
xtl, crystallite

REFERENCES

1. Bauchop T. Rumen anaerobic fungi of cattle and sheep. Appl Environ Microbiol 1979; 38:148–158.
2. Bauchop T. Scanning electron microscopy in the study of microbial digestion of plant fragments in the gut. In: Ellwood DC, Hedger JN, Latham MJ, Lynth JM, Slater JH, eds. Contemporary Microbial Ecology. London: Academic Press, 1980:305–326.
3. Bauchop T, Mountfort DO. Cellulose fermentation by a rumen anaerobic fungus in both the absence and the presence of rumen methanogens. Appl Environ Microbiol 1981; 42:1103–1110.
4. Munn EA, Orpin CG, Hall FJ. Ultrastructural studies of the free zoospore of the rumen phycomycete *Neocallimastix frontalis*. J Gen Microbiol 1981; 125:311–323.
5. Munn EA, Greenwood CA, Orpin CG. Organization of the kinetosomes and associated structures of zoospores of the rumen chytridiomycete *Neocallimastix*. Can J Bot 1987; 65:456–465.
6. Munn EA, Orpin CG, Greenwood CA. The ultrastructure and possible relationships of four obligate anaerobic chytridiomycete fungi from the rumen of sheep. BioSystems 1988; 22:67–81.

7. Orpin CG, Munn EA. *Neocallimastix patriciarum*: a new member of the Neo-callimasticaceae inhabiting the rumen of the sheep. Trans Br Mycol Soc 1986; 86:178–181.
8. Heath IB, Bauchop T, Skipp RA. Assignment of the rumen anaerobe *Neocallimastix frontalis* to the Spizellomycetales (Chytridiomycetes) on the basis of its polyflagellate zoospore ultrastructure. Can J Bot 1983; 61:295–307.
9. Webb J, Theodorou MK. A rumen anaerobic fungus of the genus *Neocallimastix*: ultrastructure of the polyflagellate zoospore and young thallus. BioSystems 1988; 21:393–401.
10. Webb J, Theodorou MK. *Neocallimastix hurleyensis* sp. nov., an anaerobic fungus from the ovine rumen. Can J Bot 1991; 69:1220–1224.
11. Gold JJ, Heath IB, Bauchop T. Ultrastructural description of a new chytrid genus of caecum anaerobe *Caecomyces equi* gen. nov., sp. nov., assigned to the Neocallimasticaceae. BioSystems 1988; 21:403–415.
12. Li J, Heath IB, Bauchop T. *Piromyces mae* and *Piromyces dumbonica*, two new species of multiflagellate anaerobic chytridiomycete fungi from the hind-gut of the horse and elephant. Can J Bot 1990; 68:1021–1033.
13. Munn EA, Orpin CG, Greenwood CA, Frith P. Ultrastructural characteristics of *Piromonas equi* (sp. nov.), an anaerobic zoosporic fungus isolated from horse caecum. Mycol Res. Submitted.
14. Barr DJS, Hadland-Hartmann VE. The flagellar apparatus in the Chytridiales. Can J Bot 1978; 56:887–900.
15. Barr DJS. An outline for the classification of the Chytridiales, and for a new order, the Spizellomycetales. Can J Bot 1980; 58:2380–2394.
16. Barr DJS. The zoosporic grouping of plant pathogens. In: Buczacki ST, ed. Zoosporic Plant Pathogens. A Modern Perspective. London: Academic Press, 1983:43–83.
17. Barr DJS. How modern systematics relate to the rumen fungi. BioSystems 1988; 21:351–356.
18. Hardham AR, Suzaki E, Perkin JL. Monoclonal antibodies to isolate-, species-, and genus-specific components on the surface of zoospores and cysts of the fungus *Phytophthora cinnamomi*. Can J Bot 1986; 64:311–321.
19. Yarlett N, Orpin CG, Munn EA, et al. Evidence for hydrogenosomes in the rumen fungus *Neocallimastix patriciarum*. Biochem J 1986; 236:729–739.
20. Heath IB, Kaminskyj SGW, Bauchop T. Basal body loss during fungal zoospore encystment: evidence against centriole autonomy. J Cell Sci 1986; 83:135–140.
21. Heath IB. *Neocallimastix*, its phylogenetic position and the value of the characters used to put it there. Protistologica 1983; 19:463.
22. Heath IB, Bauchop T. Mitosis and the phylogeny of the genus *Neocallimastix*. Can J Bot 1985; 63:1595–1604.
23. Orpin CG. Studies on the rumen flagellate *Sphaeromonas communis*. J Gen Microbiol 1976; 94:270–280.
24. Orpin CG. The occurrence of chitin in the cell walls of the rumen organisms *Neocallimastix frontalis, Piromonas communis* and *Sphaeromonas communis*. J Gen Microbiol 1977; 99:215 218.

25. Orpin CG. On the induction of zoosporogenesis in the rumen phycomycetes *Neocallimastix frontalis, Piromonas communis* and *Sphaeromonas communis.* J Gen Microbiol 1977; 101:181–189.
26. Orpin CG, Munn EA. *Piromyces communis* sp. nov., *P. lethargica* sp. nov. and *Sphaeromonas* gen. nov. *communis* sp. nov., new species of anaerobic fungi from the rumen of the sheep. Mycol Res. Submitted.
27. Heath IB. The behaviour of kinetochores during mitosis in the fungus *Saprolegnia ferax.* J Cell Biol 1980; 84:531–546.
28. Heath IB. Variant mitoses in lower eukaryotes. Indicators of the evolution of mitosis? Int Rev Cytol 1980; 64:1–80.
29. Lichtwardt RW. Trichomycetes. In: Ainsworth GC, Sparrow FK, Sussman SS, eds. The fungi, an advanced treatise Vol. IVB taxanomic review with keys: basidiomycetes and lower fungi. London: Academic Press, 1973:237–243.
30. Breton A, Bernalier A, Dusser M, et al. *Anaeromyces mucronatus* nov. gen., nov. sp. A new strictly anaerobic rumen fungus with polycentric thallus. FEMS Microbiol Lett 1990; 70:177–182.
31. Orpin CG. Studies on the rumen flagellate *Neocallimastix frontalis.* J Gen Microbiol 1975; 91:249–262.

3

Nutrition and Survival of Anaerobic Fungi

MICHAEL K. THEODOROU and DAVID R. DAVIES

Institute of Grassland and Environmental Research
Plas Gogerddan, Aberystwyth
Dyfed, Wales

COLIN G. ORPIN*

Division of Tropical Crops and Pastures
CSIRO
Brisbane, Queensland, Australia

I. INTRODUCTION

Evidence has accumulated since the mid-1970s that leaves little doubt about the existence and involvement of anaerobic fungi in the digestive tract of ruminants and other large mammalian herbivores. These microorganisms are uniquely adapted to survival in the rumen. Hence, they have a fermentative, mixed-acid metabolism and possess hydrogenosomes similar to those found in the anaerobic protozoa (1,2). Moreover, the anaerobic fungi are highly fibrolytic microorganisms, producing a wide range of cell-bound and cell-free glycolytic, cellulolytic, and hemicellulolytic enzymes (3,4,5). These enzymes are able to digest the major structural carbohydrates of plant cell-walls and permit the fungus to grow on a number of plant polysaccharides (3,6). Although the extent of their activity in ruminants has yet to be determined, it is generally acknowledged that the anaerobic fungi contribute, along with anaerobic bacteria and protozoa, to the dissolution of plant biomass in the rumen. In ruminants fed on fibrous diets, a substantial proportion of the plant fragments that enter the rumen are rapidly and extensively colonized by anaerobic fungi (7,8). Anaerobic fungi may therefore participate in the initial colonization of plant cell-walls and further assist in ruminal cel-

*Present affiliation: Consultant in Agricultural Biotechnology, Cambridge, England

lulolysis by increasing the accessibility of plant biomass to invasion by other microorganisms (6,9).

Most of the intensively studied species of anaerobic fungi (species of *Neocallimastix*, *Piromyces*, and *Caecomyces*) have been obtained from the rumen, but similar fungi can be found in ruminant feces (Table 1, Fig. 1). Although their presence in feces has been known for some time, anaerobic fungi isolated from feces have received relatively little attention in the literature (10–16). This is perhaps surprising, in view of their anaerobic status, but is probably related to the fact that techniques for enumeration of anaerobic fungi in feces were not available until comparatively recently (17). The population density of anaerobic fungi in feces, which can be numerically equivalent to that of anaerobic fungi in the rumen, was therefore not appreciated. While data on anaerobic fungi in habitats other than the rumen are limited, their presence in feces implies that they are common members of the gut microflora of many herbivorous animals (Table 1).

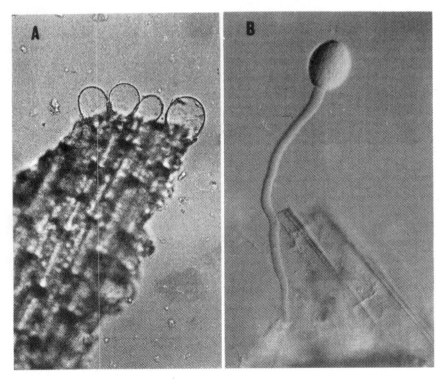

Figure 1   Anaerobic fungi of the type found in the rumen can also be isolated from feces. (A) *Caecomyces*-like fungi from cattle feces growing on wheat straw in batch culture (note the spherical structures). (B) *Neocallimastix*- or *Piromyces*-like fungus from cattle feces growing on wheat straw in batch cultures.

Table 1 Distribution of Anaerobic Fungi in Herbivores

Animal		Anaerobic fungi isolated (or observed)
Common name	Specific name	
African elephant	*Loxodonta africana*	Feces
Asian elephant	*Elephas maximus*	Feces
Arabian oryx	*Oryx leucoryx*	Feces
Bactrian camel	*Camelus bactrianus*	Feces
Blue duiker	*Cephalophus monticola*	Rumen, cecum
Bongo	*Taurotragus euryceros*	Feces
Chinese water deer	*Hydropotes inermis*	Feces
Common zebra	*Equus burchelli*	Feces
Domestic cattle	*Bos* spp.	Digestive tract, feces
Domestic goat	*Capra* spp.	Rumen, feces
Domestic sheep	*Ovis* spp.	Digestive tract, feces
Gaur	*Bos gaurus*	Feces
Greater kudu	*Tragelaphus strepsiceros*	Feces
Gray kangaroo	*Macropus giganticus*	Forestomach
Horse	*Equus caballus*	Feces, cecum
Impala	*Aepyceros melampus*	Rumen
Indian rhinocerus	*Rhinoceros unicornis*	Feces
Llama species	*Lama glama*	Feces
	Lama guanicoe	Feces
	Lama pacos	Feces
Mara	*Dolichotis patagonum*	Feces
Musk ox	*Ovibos moschatus*	Rumen
Red deer	*Cervus elaphus*	Rumen
Redneck wallaby	*Macropus rufogriseus*	(Forestomach)
Reindeer	*Rangifer tarandus*	Rumen
Black rhinoceros	*Diceros bicornis*	Feces
Roan antelope	*Hippotragus equinus*	Feces
Swamp wallaby	*Wallabia bicolor*	(Forestomach)
Vicuna	*Vicugna vicugna*	Feces
Wallaroo	*Macropus robustus*	(Forestomach)
Water buffalo	*Bubalus bubalis*	Rumen

The following references were used in table compilation: 7,8,10,11,12,15,17,34,40, 47,48,55,56,57; M. K. Theodorou and C. G. Orpin, unpublished results.

The discovery of anaerobic fungi has stimulated considerable worldwide interest, to the point where the accepted role of substrate-associated microorganisms in the digestive tract ecosystem is now in question. Their discovery has also challenged an accepted concept of mycology, the well-established dogma that fungi require oxygen to grow. Prompted by the uniqueness of the fungi themselves and by the hypothesis that they may be the initial colonizers of lignocellulosic substrates in the rumen, a number of research groups have developed the methodology for working with these organisms. To what extent anaerobic fungi participate in the hydrolysis of plant biomass in the rumen is the subject of other contributions in this textbook. In this chapter, a description of the nutrition and survival of anaerobic fungi is presented, together with an account of the media and methodology for working with these organisms. With respect to the apparent anomaly of large populations of anaerobic fungi in feces, evidence is presented for an additional stage (a survival stage) in their life cycle. It is postulated that this stage may be responsible for the presence of anaerobic fungi in feces.

II. ENUMERATION, ISOLATION, AND CULTURE

A. Anaerobic Techiques and Culture Media

Many of the culture techniques and habit-simulating media used in rumen microbiology were originated by Hungate to study anaerobic rumen bacteria (18,19). Since then, they have been modified and extended to provide the techniques commonly associated with anaerobic microbiology (20,21). With relatively few exceptions, these procedures, together with recently developed anaerobic glove-box and petri dish techniques (22,23), are now used to isolate, culture, and study the anaerobic fungi.

Anaerobic fungi are generally grown in small batch cultures (closed cultures) of 7- to 100-ml amounts in thick-walled glass tubes, serum tubes, or serum bottles, which are often sealed with butyl rubber stoppers and aluminum crimp seals (Fig. 2). Procedures are also available for the cultivation of anaerobic fungi in larger batch fermenters of 10 to 20 l in volume (24). Culture media usually inoculated with zoospore suspensions and grown without agitation in carbon-limited media on soluble (glucose, xylose, cellobiose) or insoluble (cellulose, wheat straw) substrates. In studies with larger batch fermenters, colonized particles have been used as the source of the fungus, thus increasing the potency of the inoculum and decreasing the lag time in the fermenter (24). Culture media incorporating plant tissues as a carbon source are valuable for ensuring that the anaerobic fungi do not lose their ability to ferment plant structural carbohydrates. The headspace gas above culture media consists of either 100% CO_2 or a mixture of 70–80% CO_2:20–30% N_2. In cocultures, which consist of an aerobic fungus and a methanogen, a

Figure 2 Cultures of *Neocallimastix hurleyensis* growing on wheat straw in serum bottles containing 100 ml of defined liquid medium (Medium B). The gas phase in the headspace is CO_2, and culture bottles are sealed with butyl rubber stoppers and aluminum crimp seals.

small amount of H_2 (ca. 5%) may also be included in the headspace to stimulate methane production. The incubation temperature for fungal cultures, at 39°C, is equal to that of the rumen, and incubation times range from 2 to 10 days, depending on the maintenance procedure or experiment in progress. Even though anaerobic fungi persist and survive in an open (continuous) ecosystem, relatively few studies have been performed in which anaerobic fungi have been grown in continuous culture systems or chemostats.

The composition of culture media used to grow anaerobic fungi is similar to that used for cultivation of rumen bacteria. Almost all media contain rumen fluid and are therefore described as complex and undefined. These media are well buffered to a pH of 6.5–6.8 and may be solidified with 0.8–1.5% agar before use. They contain resazurin as a redox indicator, phosphate and/or bicarbonate buffers, micro- and macrominerals, organic and/or inorganic nitrogen sources, vitamins, and the chemical reducing agents sodium sulfide and/or L-cysteine hydrochloride (24). Some culture media are prereduced (boiled and gassed with CO_2) before autoclaving and addition of chemical reducing agents, whereas others incorporate the reducing agents before autoclaving. Reducing agents and anaerobic techniques are essential during media preparation, to decrease the redox potential to that required for the cultivation of strict anaerobes. One or more of the antibacterial antibiotics ampicillin, penicillin, streptomycin, and chloramphenicol can be incorporated into culture media, particularly during enumertion and isolation of anaerobic fungi from digesta and feces. Two media lacking rumen fluid are available for more definitive studies with the anaerobic fungi (23,25). One of these (Medium B) was originally developed as a general-purpose medium for the cultivation of rumen bacteria (26). However, it was found to be suitable for the anaerobic fungi, and by omitting yeast extract and peptone, a completely defined medium was provided for growth studies (3,23,37,38).

B. Enumeration Procedures

Although zoospores in rumen fluid have been enumerated (29,30), the population density of substrate-associated thalli in digesta contents remains unknown. This is because of the difficulty of quantifying particle-attached fungal biomass in a mixed-population ecosystem. Nevertheless, because of the obvious differences in biomass content of zoospores and thalli, it is reasonable to assume that most of the fungal biomass in the rumen is in the form of digesta-associated thalli (31). Procedures for enumeration of free-swimming zoospores are relatively straightforward. Quantification is achieved either by making direct microscopic counts of zoospores or by making viable counts of colonies developing from zoospores in agar-containing roll-tubes (29,30). The population density of zoospores in the rumen of sheep, as measured by either of these two methods, is consistently within the range of 10^3–10^5 ml^{-1} (29,30). A less-used enumeration technique, which enables a comparison of thallus populations, is to count the number of zoosporangia associated with leaf blades or agar strips after in vitro or in sacco incubation in culture medium or rumen fluid (32,33). Although these procedures can be used with anaerobic fungi in rumen digesta or batch culture, they are inappropriate for fecal enumerations, where zoospores are absent and the fungi are likely to be inactive or dormant (14). These considerations led to the development of an enumeration procedure that is not constrained by an absence of zoospores and inactivity of the fungi in the sample to be enumerated (17). This method, which is shown diagrammatically in Fig. 3, uses a most-probable-number (MPN) technique to enumerte anaerobic fungi as thallus forming units (TFU). Enumeration of anaerobic fungi as TFU is analogous to enumeration of bacteria as colony forming units (CFU). In the same way as a CFU can be defined as a bacterium or collection of bacteria (rods in chains, cocci in tetrads, clusters, or chains) with the ability to produce a colony in culture, a TFU may be defined as a zoospore or collection of zoospores (inside a zoosporangium) with the ability to produce a fungal thallus in culture. The method does not distinguish between stages of the fungal life cycle, nor can it determine the numbers of thalli associated with each colonized particle. However, populations can be enumerated relative to their ability to digest

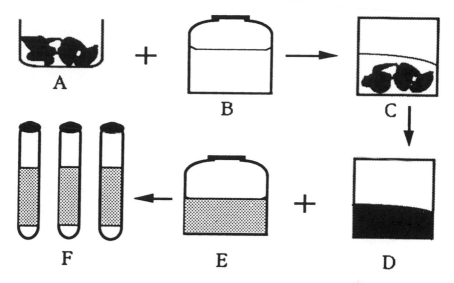

Figure 3 Diagrammatic representation of the most-probable-numbers procedure for enumeration of anaerobic fungi as thallus-forming units. The sample (A) is placed with an anaerobic medium (B) in a bag (C) and pummelled to obtain a homogeneous suspension (D). A dilution series is prepared in bottles (E), and appropriate dilutions are used to inoculate tubes containing wheat straw (F). These tubes are incubated and scored for the presence or absence of anaerobic fungi. Populations are quantified as thallus-forming units using statistical tables.

plant cell-walls. Although the method provides a minimal enumeration, with respect to the total number of viable cells in a population, it has practical merit in that it can be used to compare populations of anaerobic fungi in different habitats (15,17,34).

C. Isolation Procedures

Anaerobic fungi were first isolated from the ovine rumen by placing digesta solids on top of an antibiotic-containing sloppy-agar medium in anaerobic culture tubes. After incubation, the upper portion of the culture was discarded, while the lower portion, which contained migrating zoospores, was used to inoculate tubes of fresh sloppy-agar medium. By repeating this process several times, anaerobic fungi free from contaminating bacteria were obtained (35). This procedure was used to isolate *Neocallimastix patriciarum*, *Piromyces communis*, and *Sphaeromonas communis*, all of which became the subject of extensive research by Orpin and his co-workers. An antibiotic-containing sloppy-agar medium and enrichment culture, to increase the size of the initial fungal population, was used to isolate *N. frontalis* from the ovine rumen (36,37). In this technique, axenic cultures were obtained by using a syringe needle to transfer fungal thalli to fresh medium. A more elaborate procedure than those described above was used to obtain *N. hurleyensis* from the ovine rumen (23,38). Enrichment cultures were prepared by incubating rumen digesta in a liquid medium that contained antibiotics and milled barley straw. After incubation, these cultures were transferred to an anaerobic glove box, where the colonized straw particles were placed on petri dishes containing cellulose-overlay agar medium and antibiotics. After incubation, small plugs were removed from the margins of colonies developing on petri dishes and transferred to an antibiotic-free liquid medium that contained glucose. After incubation of antibiotic-free cultures, those contaminated with bacteria were discarded, whereas those containing fungi free from bacteria were retained.

With gradual improvements in isolation procedure and the realization that anaerobic fungi can be obtained from sources other than the rumen, it has become relatively easy to obtain these microorganisms, particularly from feces (14,17). The roll-tube technique offers perhaps the most straightforward means of obtaining cultures of anaerobic fungi. This method was originally developed to enumerate and isolate cellulolytic bacteria from the rumen (19). However, it can also be used to isolate anaerobic fungi, not only from rumen fluid, but also from feces and digesta from other parts of the ruminant digestive tract (14,29). The method involves preparation of a dilution series of the sample in an anaerobic buffer solution. Appropriate dilutions are then mixed with an antibiotic-containing molten-agar medium from which roll-tubes are prepared. Isolated fungal colonies are obtained after incubation of the roll-tubes and axenic cultures isolated by passage through successive roll-tubes (29).

D. Culture Maintenance

Anaerobic fungi grown in liquid medium require subculturing at frequent intervals (2–7 days) in order to maintain culture viability (14). Cultures grown on soluble carbohydrates require subculturing more frequently, at 24- to 48-h intervals, whereas those grown on plant tissue can remain viable for up to 7 days, thus simplifying routine culture maintenance. However, one report suggests that anaerobic fungi can survive for considerable periods, of up to 7 months, in agar-containing roll-tubes without the need for frequent subculture (29). Although the mechanism of long-term survival in these cultures was not explained, it is possible that the conditions employed somehow induced the formation of survival structures (perhaps cysts or resistant zoosporangia). Although it is not yet possible to induce the formation of these survival structures in liquid cultures, further research may elucidate their mechanism of formation and thus negate the current requirement for repeated subculture to maintain culture viability. Cryopreservation techniques using glycerol and/or dimethyl sulfoxide as a cryoprotectant and involving storage at $-70°C$ or in liquid nitrogen can be used for the long term storage of anaerobic fungi (39).

III. NUTRITIONAL CHARACTERISTICS

Of all the anaerobic fungi grown in the laboratory to date, only one strain has been grown in a fully defined minimal medium: this was *Neocallimastix patriciarum* strain H8 (25). *Neocallimastix hurleyensis*, *N. patriciarum*, and other strains of *Neocallimastix* and *Piromyces* isolates from a number of sources have been grown in complex defined media lacking rumen fluid (23, 40). When grown in the minimal medium, on cellobiose in a bicarbonate-CO_2 buffered salts solution (25), *N. patriciarum* H8 was supplied with only a source of heme; D-biotin; thiamin; ammonium ions; and L-cysteine as a sulfur source and reductant. Growth in this medium was poor, but it was markedly stimulated by a variety of amino acids, straight- and short-chain fatty acids, low concentrations of certain long-chain fatty acids, and a number of vitamins. Sulfate could not be used as sulfur source. Despite attempts to grow other strains of *Neocallimastix* and other species of anaerobic fungi in this medium, it was inadequate to sustain growth.

Virtually all known anaerobic fungi will utilize a range of mono-, di-, and polysaccharides for growth (1,3,4,5,35,36), including most of the structural plant cell-wall polysaccharides including cellulose and xylan. Many of these compounds are continuously available in the alimentary tracts of herbivores, and it should be rare for the anaerobic fungi to be deprived of metabolizable carbon sources. Their carbon and energy requirements apparently cannot be supplied adequately by proteins or amino acids.

The requirements for D-biotin and thiamin indicate similarities with the nutritional requirements of other chytrid fungi (37) and provide information confirming the close taxonomic relationship with the aerobic chytridiomycetes. The heme requirement can be met by the provision of a number of iron-containing porphyrins, but not porphyrin precursors (25), suggesting that *N. patriciarum* H8 cannot insert the iron into the porphyrin ring. Hemes also appear to act as a signal for the rumen fungi when fresh food enters the rumen. Although hemes are present at all times in the rumen contents, differentiation of the sporangia and release of zoospores from rumen anaerobic fungi can be triggered by the introduction of heme into the rumen (41). Regulation of zoosporogenesis is more complicated than this suggests, since it could not be reproduced routinely in vitro (25).

Nitrogen can be supplied to *N. patriciarum* H8 either as ammonium ions or as amino acid nitrogen (25). Growth with ammonium ions was poor compared with that on amino acids (25), tested separately and in combination. Maximum growth stimulation was achieved with glutamate, serine, and methionine. The omission of aromatic amino acids from complete amino acid mixtures resulted in a decrease in growth compared with the control, suggesting that the organism may not adequately synthesize sufficient aromatic amino acids for optimal growth.

Nitrogen could not be supplied as nucleotide bases or nitrate. Some growth was recorded wth bovine serum albumin as nitrogen source, probably because *Neocallimastix* spp. are known to be proteolytic (42) and can utilize any released amino acids, and perhaps peptides, for growth.

Sulfur can be utilized by *N. patriciarum* H8 when supplied as sulfide, L-cysteine, or methionine. Sulfate- or sulfur-containing reducing agents such as thioglycollate, dithiothreitol, and 2-mercaptoethanol were not utilized (25). The amino acid methioine, however, acted as sulfur source only when in the presence of a reductant, such as 2-mercaptoethanol. All these forms of sulfur would be generated by proteolysis and sulfate reduction in the rumen. The low redox potential in the rumen would ensure a continuous supply of sulfide, L-cysteine, and methionine to ensure the survival of the rumen fungi.

IV. SURVIVAL AND TRANSFER

A. Survival in the Rumen and in Batch Culture

The life cycle of anaerobic fungi in the rumen is reported to consist of two stages, in which motile zoospores in the rumen fluid alternate with digesta-associated fungal thalli (27,43,44,45). The life cycle lasts about 24–32 h in vitro (27), although under appropriate conditions zoosporogenesis can occur as early as 8 h after encystment (30,31). The life cycle of anaerobic fungi in grazing ruminants may therefore be only 8 h in duration. Zoospores can remain motile in rumen fluid for several hours before they settle and encyst on plant fragments, or they may encyst within minutes of release from the zoosporangium (31). A chemotactic response to soluble sugars has been demonstrated with zoospores of *Neocallimastix* spp., and this may assist in the location of freshly ingested plant fragments (46). Encysted zoospores quickly germinate to produce a fungal thallus, which is attached to plant fragments and consists of a rhizoidal system with one (monocentric species) or more (polycentric species) zoosporangia. The rhizoids may be highly branched and tapering, as in *Neocallimastix*, *Piromyces*, and *Orpinomyces* spp., or may consist of one or more spherical rhizoidal structures, as in *Caecomyces* spp. (27,44,45,47).

The life cycle of *N. hurleyensis*, which may be regarded as typical of the monocentric fungi, has been followed by time lapse photomicroscopy of thalli in role-tube cultures (27). In a typical experiment, the encysted zoospore germinated to form an aseptate main rhizoid that increased in length exponentially at a specific growth rate of $0.28 \ h^{-1}$; growth decelerated after 6.5 h, and no further extension occurred after 9.5 h. Growth of the main rhizoid provided a convenient indicator of the growth of the whole rhizoidal system. By contrast, the zoosporangium that developed on top of the main rhizoid increased in volume at a specific growth rate of $0.44 \ h^{-1}$; growth decelerated after 14 h, and no further increase occurred after 20 h. Thus, the zoosporangium grew faster than the main rhizoid and continued to grow after extension of the rhizoidal system was complete. Cessation of growth of the zoosporangium and maturation of the thallus was associated with the formation of a septum between its base and the top of the main rhizoid. This mode of growth is typical of the monocentric chytrids and consistent with the hypothesis that cytoplasm produced in the rhizoidal system contributes to the growth of the zoosporangium. Zoospores (ca. 88) that developed inside the zoosporangium were finally released from the mature thallus through a single pore that formed in the zoosporangial wall opposite the main rhizoid. Zoospore release in other anaerobic fungi is reported to occur either through several pores in the zoosporangial wall or by nonlocalized dissolution of the entire zoosporangium (37,44). Some evidence suggests that zoospore liberation in the rumen may be induced by water-soluble hemes or other components that enter with the diet (30,41). Following zoospore release in monocentric fungi, the remaining thallus autolyzes without further development (27).

Research suggests that in monocentric fungi nuclei are absent from the rhizoidal system but are present and multiply in the developing zoosporangium (27). These nuclei are destined for liberation within the mature zoospores, and thus the process of septation and zoosporulation will result in the production of an anucleate vegetative thallus without the capacity for further development. This mode of growth, in which the vegetative life cycle is of finite duration, is not unusual in the chytrids. It is of significance in rumen fungi, however, because it dictates that both zoospores and thalli are required for the continued production of fungal biomass. The situation regarding the life cycle and mode of biomass production of polycentric anaerobic fungi is less clear and awaits more detailed studies similar to those performed with *N. hurleyensis* and other monocentric fungi.

The rumen is an open (continuous) ecosystem with the combined attributes of absorption of fermentation end products and passage of digesta solids to ensure a relatively constant environment (stable-state) and prevent the build-up of toxic compounds or development of adverse physiological conditions. In this environment, both zoospores and thalli can coexist in a state of dynamic equilibrium. Thus, in the absence of substantial environmental perturbations (such as a significant change in the nature of the diet), the fungal population will remain viable indefinitely. The size of the zoospore and thallus populations is likely to be dependent on a number of factors, including substrate type and availability, the respective swimming and encystment times of zoospore and thallus populations, the maturation time of fungal thalli, the yield of zoospores from mature thalli, and the flow properties of liquid and solids from the rumen. Incorporating some of these factors into the framework of a mathematical model has provided an indirect means of estimating the thallus population in the rumen (31; and Chapter 10 in this book).

Unlike anaerobic fungi in rumen digesta, those growing in batch culture cannot be maintained indefinitely without repeated subculturing. Loss of culture viability and autolysis of fungal biomass in cultures grown on soluble sugars has been shown to coincide with the depletion of the carbon source (28). By contrast, in cultures grown on insoluble substrates, which generally employ higher substrate concentrations, end-product inhibition and/or the development of adverse physiological conditions are more likely to be responsible for the loss of culture viability (1,3). In fungi that have a determinate life span, any condition that interrupts the ability of the fungus to produce zoospores will ultimately lead to the death of the culture. The outcome of this phenomenon was observed in batch cultures of *N. hurleyensis* where zoospores were enumerated during the fermentation of wheat straw (17). After an initial increase in zoospore numbers, the zoospore population declined rapidly, while the fungus continued to remove dry matter from wheat straw (Fig. 4). Presumably, changing conditions in the culture medium became toxic to zoospores or inhibited zoospore release before affecting the fibrolytic activity of the fungal thalli. Irrespective of the cause, however, the effect of a declining zoospore population ultimately led to the loss of culture viability.

Most rumen microorganisms can grow only under anaerobic conditions and over a relatively restricted temperature range. The effects of aeration and changing temperature on anaerobic fungi have been investigated, and results generally show that both zoospores and thalli are unable to survive under conditions that are likely to prevail outside the animal (13,28,48). Although anaerobic fungi are reported to be obligate anaerobes, they do demonstrate a limited degree of tolerance to air. *N. hurleyensis*, for example, was sucessfully isolated from stationary-phase cultures 14 h after they had been aerated for 1 min and the liquid culture subsequently kept under air at 39°C (11,13). Thus, although anaerobic fungi are unable grow in the presence of air, they are sufficiently tolerant to survive in air for a few hours. In inves-

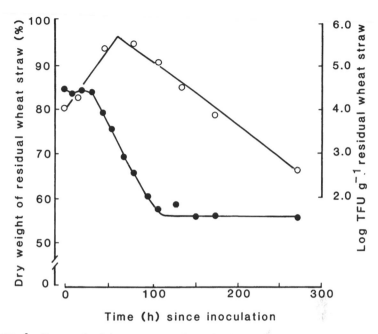

Figure 4 Removal of dry matter and production of zoospores during fermentation of wheat straw by batch cultures of *Neocallimastix hurleyensis*. Symbols: ●, wheat straw residue; ○, zoospores. (From Ref. 17.)

tigations with *N. hurleyensis*, the optimum temperature for zoospore development was between 36 and 39°C, and zoospores did not germinate above 42°C or below 25°C (11,13). Similarly, immature thalli of *N. hurleyensis* failed to grow and produce zoospores when transferred from 39°C to temperatures below 30°C (11,13). Although *N. hurleyensis* will not grow outside this narrow temperature range, it may be that zoospores and/or thalli can remain viable at elevated or reduced temperatures. To investigate this, zoospores and immature thalli were moved from their optimal incubation temperature to 20°C for various periods and then reincubated at 39°C and tested for viability. Results showed that the viability of both stages decreased rapidly, reinforcing the hypothesis that zoospores and thalli may not be able to survive at reduced temperatures outside the rumen (11,13,28).

B. Survival in the Hindgut and Feces

In addition to the rumen, anaerobic fungi have now been isolated from other parts of the ruminant digestive tract and from feces. They have also been isolated from the horse cecum, the forestomach of the kangaroo, and the feces of numerous other herbivorous mammals, including macropod marsupials (11,48; Table 1; M. K. Theodorou and C. G. Orpin, unpublished results). Moreover, anaerobic fungi identical in appearance to rumen chytridiomycetes have been isolated from cattle and sheep feces that were dried and stored in air at ambient temperature for up to 10 months (11,17). Populations of anaerobic fungi in feces can be substantial (10^4–10^5 TFU of anaerobic fungi g^{-1} fecal dry matter) (17), and they decline very slowly after drying (Fig. 5). By culturing anaerobic fungi from sun-baked and dry feces collected from the dung heaps of cattle and sheep in Ethiopia, their survival under more natural field conditions has also been demonstrated (11).

A major difference between fungal populations in the rumen and those present in the hindgut and in feces is the ability of the latter two to survive for long periods under conditions of reduced temperature, desiccation, and exposure to air (11,14–17). Despite numerous attempts, significant populations of anaerobic fungi have not been obtained from dried rumen digesta (15,16), even when digesta solids were washed and dried under an-

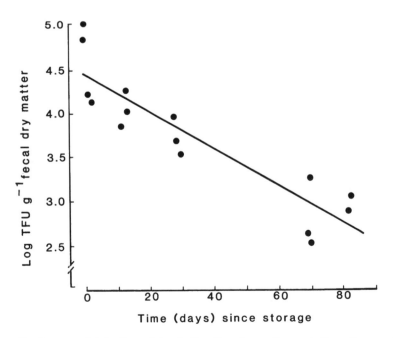

Figure 5 Exponential decline of the thallus-forming unit population of anaerobic fungi in cattle feces during 90 days of storage in air at ambient temperature. (From Ref. 14).

aerobic conditions (D. R. Davies, M. K. Theodorou, C. G. Orpin, and A.P.J. Trinci, unpublished results). In one study, where samples of rumen digesta were taken from different regions of the rumen, including an area close to the omasal orifice, all samples were found to be absent of viable anaerobic fungi after drying (14). Rumen digesta have also been mixed and dried with feces, whereupon a proportion of the population, similar to that found in feces alone, survived the drying process (14). When, however, the rumen digesta were mixed and dried with sterile feces, anaerobic fungi in rumen digesta failed to survive the drying process (14). From these and similar experiments (15,17), it was concluded that the lack of anaerobic fungi in dried rumen digesta was not due to the concentration of substances such as the volatile fatty acids to toxic levels during drying, but was more probably related to the form of the fungi.

Each organ of the ruminant digestive tract has now been examined for populations of anaerobic fungi (15). In recent studies, digesta from the rumen (1.17×10^5 TFU g^{-1} dry matter [DM]), omasum (1.82×10^5 TFU g^{-1} DM), abomasum (5.75×10^3 TFU g^{-1} DM), small intestine (2.18×10^2 TFU g^{-1} DM), cecum (2.88×10^4 TFU g^{-1} DM), and large intestine (4.90×10^4 TFU g^{-1} DM) were all found to contain fungal populations; values in parentheses represent the size of the digesta-associated TFU populations g^{-1} in each organ in young steers (15). Upon air-drying these samples, the fungal population in rumen digesta declined to zero, whereas a proportion of the fungi present in each of the other organs survived the drying process (15). Moreover, the viability of fungal populations upon air-drying increased with distance down the alimentary tract (Fig. 6).

In order to account for these observations, the accepted life cycle for anaerobic fungi has been amended as shown diagrammatically in Fig. 7. Cycle A is the generally accepted vegetative cycle (8,27,35). An alternative route is indicated by cycle B, which is activated when conditions are harsh enough to inhibit vegetative growth. Accordingly, it has been speculated that a resting structure is formed postruminally that is capable of germination, either within the hindgut or within another host, once conditions become favorable again. As conditions for active growth decline, with distance down the tract, the equilibrium between the germination/growth cycle (A) and the dormancy cycle (B) appears to change in favor of dormancy (Figs. 6,7).

Two previous reports hinted that anaerobic fungi produce survival structures. Although survival was not demonstrated, thick-walled zoosporangia were observed in cecal contents of the horse (48). Perhaps of more significance was the report that anaerobic fungi could be maintained in sisal roll-rubes for up to 7 months without the need for transfer (29). More evidence was provided recently by Wubah et al. (49). These authors reported the presence of resistant zoosporangia in batch cultures. These structures possessed

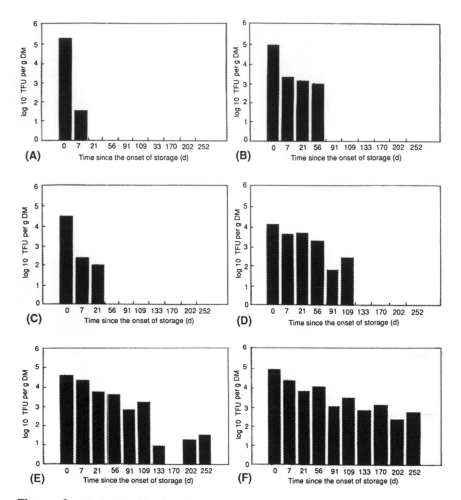

Figure 6 Fungal thallus-forming unit populations in fresh and dried digesta and feces from mature steers, dried and stored in air for up to 252 days at ambient room temperature. Each value represents the mean of three observations. The most-probable-numbers enumeration used 10-fold dilutions and three replicates per dilution. A: reticulo-rumen. B: omasum. C: abomasum. D: cecum. E: large intestine. F: feces. Data for the small intestine are not presented, as digesta from this organ failed to support a viable fungal population upon drying in air. (From Ref. 15.)

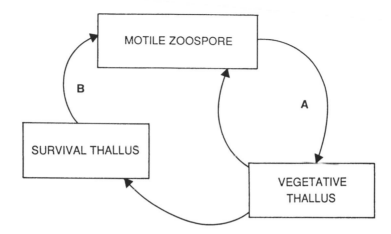

Figure 7 A diagrammatic representation of the complete life cycle of anaerobic fungi. Cycle A is the vegetative cycle purported to occur in the rumen. Cycle B is the survival cycle, which occurs when conditions for growth become unfavorable. (From Ref. 15.)

melanin, and their DNA levels were increased in comparison with normal vegetative sporangia. However, although attractive candidates for survival structures, they failed to germinate after their production in batch culture.

C. Transfer of Anaerobic Fungi Between Herbivores

Newly born ruminants lack a microbial flora and acquire it only after contact with older animals. Although anaerobic fungi are found in adult ruminants, they are not found in very young isolated, milk-fed animals (50). Licking and grooming of young animals is thought to be a major route by which bacterial and protozoal inoculation occurs (51,52). Several species of rumen bacteria have been isolated from air samples collected in a cow barn, providing evidence for transfer of these microorganisms between animals in aerosols, food, or common drinking water (53). Rumen bacteria, but not rumen protozoa, can be isolated from fecal material (18,54), thus providing an alternative route for their transfer. By isolating anaerobic fungi from both saliva and feces of sheep, it was concluded that either route may result in transfer and faunation of young animals with anaerobic fungi (12). However, as ruminants are not normally coprophagic, transfer of anaerobic fungi in feces would appear to be unlikely, although accidental contact with feces particularly by contamination of the diet might occur. Nevertheless, anaerobic fungi can be cultured routinely from feces, presumably from survival structures. these structures might be disseminated from feces to herbage in nature, thus enabling transfer of anaerobic fungi between herbivores. Such a mechanism would account for the apparent lack of host specificity of the anaerobic fungi and explain why an apparently limited range of fungal types can be obtained from a wide range of herbivores.

V. CONCLUSIONS

On the basis of current evidence, the nutritional requirements of rumen anaerobic fungi are not very demanding. While only a single species (20) has been grown on a minimal medium, defined media (24) capable of supporting the growth of *Neocallimastix* spp. have been devised that may support the growth of other anerobic chytridiomycetes. Their requirement of few growth factors supports their inclusion in the Chytridiomycetes. The determination of their precise nutritional parameters will take further study.

Techniques are now available for the routine isolation and enumeration of anaerobic fungi from the alimentary tracts of herbivores. No doubt further refinements will be made based on the establishment methods, particularly in relation to enumeration of resistant stages.

The identity of the stage resistant to oxygen and desiccation, and the occurrence of "warty spores" in the rumen, indicates that the life cycles of these organisms may be very different from the simplistic life cycle currently described. However, resting or resistant stages either may be unimportant to metabolism within the rumen, or may represent an as yet undescribed sexual phase in the life cycles of these organisms that may be crucial to their continuing existence in the rumen.

Other areas requiring examination are the longevity of the oxygen-resistant survival stage, its role in transfer of the organism to new hosts, and the role of such stages in interspecies transfers. As yet no single species of anaerobic fungus has been shown to be restricted to a single mammal, and cross-species contamination may be common but merits close examination.

REFERENCES

1. Theodorou MK, Lowe SE, Trinci APJ. Fermentative characteristics of anaerobic rumen fungi. BioSystems 1988; 21:371–376.
2. Yarlett N, Rowlands C, Yarlett NC, et al. Respiration of the hydrogenosome-containing fungus *Neocallimastix patriciarum*. Arch Microbiol 1987; 148:25–28.
3. Lowe SE, Theodorou MK, Trinci APJ. Cellulases and xylanase of an anaerobic rumen fungus grown on wheat straw, wheat straw holocellulose, cellulose, and xylan. Appl Environ Microbiol 1987; 53:1216–1223.

4. Williams AG, Orpin CG. Polysaccharide-degrading enzymes formed by three species of anaerobic rumen fungi grown on a range of carbohydrate substrates. Can J Microbiol 1987; 33:418–426.

5. Williams AG, Orpin CG. Glycoside hydrolase enzymes present in the zoospores and vegetative growth stages of the rumen fungi *Neocallimastix patriciarum, Piromonas communis,* and an unidentified isolate grown on a range of carbohydrates. Can J Microbiol 1987; 33:427–434.

6. Theodorou MK, Longland AC, Dhanoa MS, et al. Growth of *Neocallimastix* sp. strain R1 on Italian ryegrass hay: removal of neutral sugars from plant cell walls. Appl Environ Microbiol 1989; 55:1363–1367.

7. Bauchop T. Rumen anaerobic fungi of cattle and sheep. Appl Environ Microbiol 1979; 38:148–158.

8. Bauchop T. The rumen anaerobic fungi: colonizers of plant fibre. Ann Rech Vet 1979; 10:246–248.

9. Akin DE, Bornemann WS, Lyon CE. Degradation of leaf blades and stems by monocentric and polycentric isolates of ruminal fungi. Anim Feed Sci Technol 1990; 31:205–221.

10. Bauchop T. The gut anaerobic fungi: colonizers of plant fibre. In: Wallace G, Bell L, eds. Fibre in Human and Animal Nutrition. Wellington, New Zealand: Royal Society of New Zealand, 1983:143–148.

11. Milne A, Theodorou MK, Jordan MGC, et al. Survival of anaerobic fungi in faeces, in saliva, and in pure culture. Exp Mycol 1989; 13:27–37.

12. Lowe SE, Theodorou MK, Trinci APJ. Isolation of anaerobic rumen fungi from saliva and faeces of sheep. J Gen Microbiol 1987; 133:1829–1834.

13. Trinci APJ, Lowe SE, Milne A, Theodorou MK. Growth and survival of rumen fungi. BioSystems 1988; 21:357–363.

14. Davies DR, Theodorou MK, Brooks AE, Trinci APJ. Influence of drying on the survival of anaerobic fungi in rumen digesta and faeces of cattle. FEMS Microbiol Lett 1993; 106:59–64.

15. Davies DR, Theodorou MK, Lawrence MI, Trinci APJ. Distribution of anaerobic fungi in the digestive tract of cattle and their survival in faeces. J Gen Microbiol 1993; 139:1395–1400.

16. Theodorou MK, Davies DR, Jordan MGC, et al. Survival of anaerobic fungi in rumen digesta and faeces of newly-born and adult ruminants. Mycol Res 1993; 97:1245–1252.

17. Theodorou MK, Gill M, King-Spooner C, Beever DE. Enumeration of anaerobic chytridiomycetes as thallus forming units: novel method for quantification of fibrolytic fungal populations from the digestive tract ecosystem. Appl Environ Microbiol 1990; 56:1073–1078.

18. Hungate RE. The Rumen and its Microbes. London: Academic Press, 1966.

19. Hungate RE. A roll tube method for cultivation of strict anaerobes. In: Norris JR, Ribbons DW, eds. Methods in Microbiology. Vol. 3B. London: Academic Press, 1969:117–132.

20. Bryant MP. Commentary on the Hungate technique for culture of anaerobic bacteria. Am J Clin Nutr 1972; 25:1324–1328.

21. Miller TL, Wolin MJ. A serum bottle modification of the Hungate technique for cultivating obligate anaerobes. Appl Microbiol 1974; 27:985-987.

22. Leedle JAZ, Hespell RB. Differential carbohydrate media and anaerobic replica plating techniques in delineating carbohydrate-utilizing subgroups in rumen bacteria. Appl Environ Microbiol 1980; 39:709-719.

23. Lowe SE, Theodorou MK, Trinci APJ, Hespell RB. Growth of anaerobic rumen fungi on defined and semi-defined media lacking rumen fluid. J Gen Microbiol 1985; 131:2225-2229.

24. Theodorou MK, Trinci APJ. Procedures for the isolation and culture of anaerobic fungi. In: Nolan JV, Leng RA, Demeyer DI, eds. The Roles of Protozoa and Fungi in Ruminant Digestion. Armidale, Australia: Penambul Books, 1989: 145-152.

25. Orpin CG, Greenwood Y. Nutritional and germination requirements of the rumen chytridiomycete *Neocallimastix patriciarum*. Trans Br Mycol Soc 1986; 86:103-109.

26. Hazlewood GP, Theodorou MK, Hutchings A, et al. Preparation and characterization of monoclonal antibodies to a *Butyrivibrio* sp. and their potential use in the identification of rumen butyrivibrios, using an enzyme-linked immunosorbent assay. J Gen Microbiol 1986; 132:43-52.

27. Lowe SE, Griffith GW, Milne A, et al. The life cycle and growth kinetics of an anaerobic rumen fungus. J Gen Microbiol 1987: 133:1815-1827.

28. Lowe SE, Theodorou MK, Trinci APJ. Growth and fermentation of an anaerobic rumen fungus on various carbon sources and effect of temperature on development. Appl Environ Microbiol 1987; 53:1210-1215.

29. Joblin KN. Isolation, enumeration and maintenance of rumen anaerobic fungi in roll tubes. Appl Environ Microbiol 1981; 42:1119-1122.

30. Orpin CG. On the induction of zoosporogenesis in the rumen phycomycetes *Neocallimastix frontalis, Piromonas communis* and *Sphaeromonas communis*. J Gen Microbiol 1977; 101:181-189.

31. France J, Theodorou MK, Davies D. Use of zoospore concentrations and life cycle parameters in determining the population of anaerobic fungi in the rumen ecosystem. J Theoret Biol 1990; 147:413-422.

32. Akin DE, Gordon GLR, Hogan JP. Rumen bacterial and fungal degradation of *Digitaria pentzi* grown with and without sulphur. Appl Environ Microbiol 1983; 46:738-748.

33. Ushida K, Tanaka H, Kojima Y. A simple *in situ* method for estimating fungal populations in the rumen. Lett Appl Microbiol 1989; 9:109-111.

34. Davies D, Theodorou MK, Trinci APJ. Anaerobic fungi in the digestive tract and faeces of growing steers: evidence for a third stage in their life cycle. 4th International Mycology Congress, Regensburg, Germany, 1990.

35. Orpin CG. Studies on the rumen flagellate *Neocallimastix frontalis*. J Gen Microbiol 1975; 91:249-262.

36. Bauchop T, Mountfort DO. Cellulose fermentation by a rumen anaerobic fungus in both the absence and the presence of rumen methanogens. Appl Environ Microbiol 1981; 42:1103-1110.

37. Heath IB, Bauchop T, Skipp RA. Assignment of the rumen anaerobe *Neocallimastix frontalis* to the Spizellomycetales (Chytridiomycetes) on the basis of its polyflagellate zoospore ultrastructure. Can J Bot 1983; 61:295–307.
38. Webb J, Theodorou MK. *Neocallimastix hurleyensis* sp. nov., an anaerobic fungus from the ovine rumen. Can J Bot 1991; 69:1220–1224.
39. Yarlett NC, Yarlett N, Orpin CG, Lloyd D. Cryopreservation of the anaerobic rumen fungus *Neocallimastix patriciarum*. Lett Appl Microbiol 1986; 3:1–3.
40. Teunissen MJ, Opden Camp HJM, Orpin CG, et al. Comparison of growth characteristics of anaerobic fungi isolated from ruminant and non-ruminant herbivores during cultivation in a novel medium. J Gen Microbiol 1991; 137: 1401–1408.
41. Orpin CG, Greenwood Y. Effects of haemo and related compounds on growth and zoosporogenesis of the rumen phycomycete *Neocallimastix frontalis* H8. J Gen Microbiol 1976; 132:2179–2185.
42. Wallace RJ, Munro CA. Influence of the rumen anaerobic fungus *Neocallimastix frontalis* on the proteolytic activity of a defined mixture of rumen bacteria grown on a solid substrate. Lett Appl Microbiol 1986; 3:23–26.
43. Orpin CG. Invasion of plant tissue in the rumen by the flagellate *Neocallimastix frontalis*. J Gen Microbiol 1977; 98:423–430.
44. Orpin CG. Studies on the rumen flagellate *Sphaeromonas communis*. J Gen Microbiol 1976; 94:270–280.
45. Orpin CG. The rumen flagellate *Piromonas communis*: its life history and invasion of plant material in the rumen. J Gen Microbiol 1977; 99:107–117.
46. Orpin CG, Bountiff L. Zoospore chemotaxis in the rumen phycomycete *Neocallimastix frontalis*. J Gen Microbiol 1978; 104:113–122.
47. Gold JJ, Heath IB, Bauchop T. Ultrastructural description of a new chytrid genus of caecum anaerobe, *Caecomyces equi* gen. nov., sp. nov., assigned to the Neocallimasticaceae. BioSystems 1988; 21:493–515.
48. Orpin CG. Isolation of cellulolytic phycomycete fungi from the caecum of the horse. J Gen Microbiol 1981; 123:187–196.
49. Wubah DA, Fuller MS, Akin DE. Resistant body formation in *Neocallimastix* sp. an anaerobic fungus from the rumen of a cow. Mycologia 1991; 83:40–47.
50. Fonty G, Gouet P, Jouany JP, Senand J. Establishment of the microflora and anaerobic fungi in the rumen of lambs. J Gen Microbiol 1987; 133:1835–1843.
51. Becker ER, Hsuing TS. The method by which ruminants acquire their fauna of infusoria and remarks concerning experiments on host specificity of protozoa. Proc Natl Acad Sci USA 1929; 15:684.
52. Eadie JM. The development of rumen microbial populations in lambs and calves under various conditions of management. J Gen Microbiol 1962; 29:563–578.
53. Mann SO. Some observations on the airborne dissemination of rumen bacteria. J Gen Microbiol 1963; 33:ix.
54. Hobson PN. Rumen microorganisms. Prog Indust Microbiol 1971; 9:42–77.
55. Ho YW, Abdullah N, Jalaludin S. Colonization of guinea grass by anaerobic rumen fungi in swamp buffalo and cattle. Anim Feed Sci Technol 1988; 22: 161–172.

56. Orpin CG. Fungi in ruminant nutrition. In: Degradation of Plant Cell Wall Material. London: Agricultural Research Council, 1981:36–47.
57. Dehority BA, Varga GA. Bacterial and fungal numbers in ruminal and caecal contents of the blue duiker (*Cephalus monticola*). App Environ Microbiol 1991; 57:469–472.

4

Fermentation Product Generation in Rumen Chytridiomycetes

NIGEL YARLETT

Pace University
New York, New York

"All animals are equal, but some are more equal than others"
George Orwell — Animal Farm

I. INTRODUCTION

The metabolism of carbohydrates by microorganisms is accomplished via a number of metabolic routes. The aerobic fungi form a homogeneous group with respect to sugar catabolism, ultimately producing CO_2 and H_2O and maintaining a redox balance via mitochondrial electron transport.

Anaerobic rumen chytridiomycetes, of which three genera have been isolated in pure culture (*Neocallimastix* spp., *Sphaeromonas* spp., and *Piromonas* spp. [1]), constitute a ubiquitous population of fungi inhabiting the rumen of wild and domestic ruminants, including cattle, sheep, and reindeer (2,3), and the hindgut of other large herbivores, including the horse, elephant, and rhinoceros (4,5,6). These fungi have a life cycle consisting of a motile flagellate zoospore phase and a nonmotile vegetative reproductive phase. Although lacking mitochondria, the Neocallimasticaceae do possess hydrogenosomes and lyososomelike structures (7). This cellular organization, characterized by the absence of mitochondria and a fermentation-based metabolism, is also found in some aerotolerant and anaerobic protozoa, including the trichomonads and rumen ciliates (8,9).

Oxidative decarboxylation of pyruvate is a key reaction of intermediary metabolism. Aerobic microorganisms exploit the high reducing power of pyruvate for the reduction of low-potential electron carriers, with the concomitant formation of an energy-rich thioester bond between coenzyme A and decarboxylated pyruvate. In aerobes this process is catalyzed by the pyruvate dehydrogenase complex, which uses nicotinamide-adenine dinucleotide (NAD) as electron acceptor and fulfills the requirements of respiratory metabolism, i.e., unidirectional transfer of electrons to NAD^+, the ultimate donor of oxidative phosphorylation. Under anaerobic conditions, NAD^+ would act as an electron trap with limited capacity and a redox potential unsuitable for easy removal of redox equivalents by the hydrogenase reaction (10).

During anaerobic fermentation, products of both higher and lower oxidation states than the substrate are formed (11). The anaerobic fungi, like many rumen and enteric bacteria, participate in mixed-acid fermentation, converting hexoses and pentoses to formate, acetate, lactate, ethanol, CO_2 and H_2 (12,13). The ratio of the end products varies with both species and growth conditions. For instance, *Neocallimastix patriciarum* produces only trace amounts of formate and ethanol (14,15), in contrast to *N. frontalis*, which produces formate and ethanol as major fermentation end products (12). The presence of the major glycolytic enzymes, together with the absence of glucose-6-phosphate dehydrogenase and the distribution of [^{14}C] in tracer studies, suggests that glycolysis is the sole mechanism of glucose metabolism in *N. patriciarum* (5,7), *N. frontalis* EB188 (16), and probably the anaerobic fungi as a whole.

The pyruvate formed by glycolysis is in turn converted to the major end products acetate, lactate, and ethanol (7,16). The formation of acetate is localized within hydrogenosomes, which function as redox organelles. The precursor substrate for this organelle is cytosolic malate, which is formed from oxaloacetate by malate dehydrogenase (7,16). The oxaloacetate may in turn be formed by the action of phosphoenolpyruvate carboxykinase, which results in the energetically favorable conservation of the high-energy phosphate bond in the form of a nucleoside triphosphate (7). Alternatively, pyruvate carboxylase is proposed as being responsible for oxaloacetate formation in *N. frontalis* EB188 (16); however, this reaction would result in the net expenditure of energy. The lack of phosphoenolpyruvate carboxykinase in *N. frontalis* EB188 may indicate that subtle differences exist between strains in this regard. However, a phosphoenolpyruvate carboxykinase has been reported from other *N. frontalis* strains (17). Decarboxylation of malate within the hydrogenosome effectively traps the substrate, as it is unlikely that hydrogenosomal pyruvate freely exchanges with the cytosolic pyruvate pool. The comments of Fallon et al. (16) are valid in that should this occur, it would have the net result of creating a futile cycle. The proposed pathway of glucose metabolism indicates that, as in many organisms adapted to anaerobic environments, phosphoenolpyruvate is the branch point for determining the fate of the end product formed (Fig. 1).

In several species of protozoa, e.g., trichomonads and the rumen ciliates, the reducing equivalents generated during fermentation are eliminated in the form of molecular hydrogen (8,9). In some other protozoal species, e.g., *Giardia lamblia* and *Entamoeba histolytica*, hydrogenase is absent, and in these anaerobes reducing equivalents are presumably removed by the re-

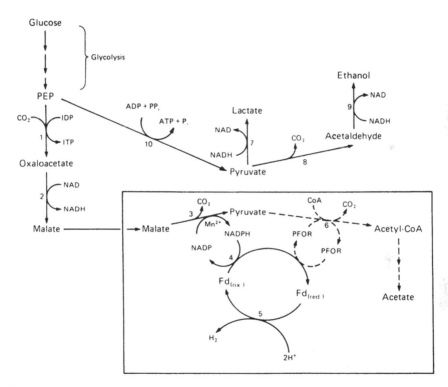

Figure 1 Metabolic transformation of glucose to acetate, lactate, and ethanol by *Neocallimastix patriciarum*. Reactions within the box are sedimentable at 10^5 g-min. 1, phosphoenolpyruvate carboxykinase; 2, malate dehydrogenase; 3, "malic" enzyme; 4, NADPH:ferredoxin oxidoreductase; 5, hydrogenase; 6, pyruvate:ferredoxin oxidoreductase; 7, lactate dehydrogenase; 8, pyruvate decarboxylase; 9, alcohol dehydrogenase; 10, pyruvate kinase. PEP = phosphoenolpyruvate; PFOR = pyruvate:ferredoxin oxidoreductase; Fd = ferredoxin. Broken lines indicate reactions suppressed by growth under high CO_2 concentrations. (Modified from Ref. 7.)

duction of acetaldehyde to ethanol by a cytosolic alcohol dehydrogenase (18,19), as occurs in trichomonads that are deficient in hydrogenase activity because of growth under metronidazole pressure (20,21).

II. LOCALIZATION OF FERMENTATIVE ENZYMES

In common with several other microorganisms, the formation of hydrogen and acetate by *N. patriciarum* is localized within hydrogenosomes. These organelles have a finely granular matrix and no apparent internal substructure (7). *Neocallimastix* hydrogenosomes appear to have a single membrane and are approximately 0.5–1.0 μM in diameter (7,22,23) (Fig. 2). Similar microbodylike organelles have been observed in *N. frontalis* and *Neocallimastix* sp. R1 (22,23,24). Hydrogenosomes in electron micrographs of thin sections of the motile zoospore stage of *Neocallimastix* appear to be clustered close to the flagella apparatus (Fig. 2) and have a close association with the kinetosomes (23,24). In some cases the association is so strong that the hydrogenosomes appear to be pulled toward the kinetosomes by an interaction with the microtubules (23). As pointed out by Heath et al. (23), the chytridiomycetes commonly cluster their mitochondria around the kinetoplast, presumably to increase the efficiency of energy transfer from the site of adenosine triphosphate (ATP) formation (25). Trichomonad hydrogenosomes are also localized close to the flagellar structure, resulting in the original description of these organelles as paraxostylar or paracostal granules (26); in addition, rumen ciliate hydrogenosomes are concentrated on the inner side of the fibrous band that separates the endoplasm and ectoplasm, in close proximity to the flagellar apparatus (9,27). Hence the intracellular localization of the organelle appears to be a consistent feature among the taxonomically very different organisms that have been shown to possess hydrogenosomes, bolstering the hypothesis that hydrogenosomes may have an important role in the provision of energy for cell motility (7,9,23). Other similarities with protozoal hydrogenosomes are the internal regions of greater electron density and the presence of protrusions from some, giving them a "dumbbell" appearance (7,23). The equilibrium density of *N. patriciarum* hydrogenosomes is about 1.20 g ml^{-1} in sucrose, 1.12 g ml^{-1} in Percoll, and 1.27–1.28 g ml^{-1} in metrizamide (7).

Hydrogenosomal enzymes produce acetyl-CoA and hydrogen by sequential decarboxylation of malate (Fig. 1). This is accomplished by "malic" enzyme and pyruvate:ferredoxin oxidoreductase and necessitates the reduction of a low-redox-potential electron-acceptor protein (ferredoxin or flavodoxin). Pyridine nucleotides cannot serve as primary electron acceptors for this class of enzyme. Hydrogenase mediates the reoxidation of the electron carrier, resulting in the formation of H_2. Coenzyme A is liberated from acetyl-CoA, forming acetate and possibly conserving the energy of the thioester

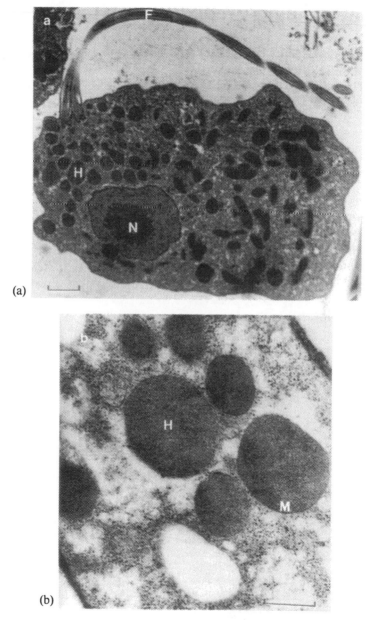

Figure 2 Electron micrographs of whole cell sections and subcellular fractions of *N. patriciarum*. Hydrogenosomes (H) in whole cells of **(a)** zoospores, where the organelles appear to be clustered around the flagella apparatus. **(b)** Rhizoids, where the hydrogenosomes appear to be randomly distributed in small groups. The organelles are about 0.5–1.00 μm in diameter and have a finely granular appearance. N = nucleus; F = flagella; M = marginal plate. Marker bars = 1 μm.

bond by thiol transfer with ATP or guanosine triphosphate (GTP) formation, as is found in the trichomonads (8) and rumen ciliates (9); however, the production of energy by this mechanism still remains to be demonstrated in the rumen chytridiomycetes.

III. FORMATION OF HYDROGEN

Two pathways of hydrogen formation have been described. Both involve pyruvate as an intermediate, but they differ in the mechanism of end-product formation. In common with the clostridia (28) and certain protozoa (9,29), *N. patriciarum* obtains electrons for proton reduction from the multistep process that is initiated by the oxidation of pyruvate to acetyl-CoA, with the concomitant reduction of an electron carrier in the form of ferredoxin or flavodoxin (Fig. 2). In coculture with rumen methanogens, the major fermentation end products of *N. frontalis* shift from acetate, lactate, formate, ethanol, CO_2, and CH_4 to predominantly acetate, CO_2, and CH_4; ethanol and lactate are greatly reduced, formate is produced in trace amounts, and hydrogen becomes undetectable (12,30,31,32). As pointed out by Bauchop and Mountfort (12), the shift in these redox balance end products is strongly suggestive that the rumen fungi possess a hydrogenase that catalyzes the production of H_2 from reduced nucleotides. The production of H_2 from reduced pyridine nucleotides is thermodynamically unfavorable at partial pressures of H_2 above 10^{-3} atm (0.1 kPa) (33). Thus the methanogens, by removing H_2 and maintaining low partial pressures, would facilitate the production of H_2 from reduced pyridine nucleotides by the fungus, resulting in a shift of redox end products from lactate and ethanol to predominantly acetate and H_2 (12). Support for the interaction of pyridine nucleotides with hydrogenase is provided by the presence in *N. patriciarum* and *N. frontalis* EB188 extracts of an NADPH (NADH):ferredoxin oxidoreductase (7,16) that can couple the transfer of electrons from reduced pyridine nucleotides to hydrogenase (Fig. 1).

When *N. patriciarum* was grown under an atmosphere high in CO_2 and in the presence of bicarbonate, pyruvate:ferredoxin oxidoreductase was suppressed, and under these conditions it is proposed that the NADPH (NADH): ferredoxin oxidoreductase activities provide the major reducing power for hydrogen formation (7). The failure of Fallon et al. (16) to detect pyruvate: ferredoxin oxidoreductase in *N. frontalis* may also be the result of the growth conditions used.

A. Hydrogenase

Hydrogenases are iron-sulfur proteins, although the iron and sulfide content, and the type of iron–sulfur cluster present, vary considerably in different species. membrane-bound hydrogenases typically contain 4Fe-4S protein, e.g., *Chromatium vinosum* and *Rhodospirillum rubrum* (34,35), or a 2[4Fe-4S] protein, e.g., *Proteus mirabilis* (36). In contrast, soluble hydrogenases are more varied in their Fe-S content; for example, the ferredoxin-dependent hydrogenase of *Clostridium pasteurianum* contains 3[4Fe-4S] clusters (37), whereas the *Methanosarcina barkeri* hydrogenase and the NAD-specific *Alcaligenes eutrophus* H16 hydrogenase contain [4Fe-4S] clusters (38,39).

The nature of the *Neocallimastix* hydrogenase has not been investigated in detail. The electron spin resonance (ESR) spectrum of *N. patriciarum* has a complex line spectrum in the oxidized state, with a predominant signal at $g = 2.05$ (where g is the absolute magnetic field position of the line of the ESR spectrum; Fig. 3). These signals are abolished by dithionite treatment (N. Yarlett, R. Cammack, and D. Lloyd, unpublished results). This is analogous to membrane-bound hydrogenases, which typically exhibit a sharp ESR signal near $g = 2.02$ at low temperature, and become silent in the reduced state (40,41). In contrast, soluble hydrogenases often exhibit complex ESR spectra in the reduced state (38,39). The signals to low field in *N. patriciarum* hydrogenosome preparations at $g = 2.48, 2.35$, and 2.22 (Fig. 3) are more prominent in the oxidized preparation, and while the nature of these signals is speculative, it is plausible that they may be due to Ni^{III}, based on the similarity in behavior of these signals to those present at $g = 2.30, 2.23$, and 2.02 in preparations of *Desulfovibrio gigas* (41) and *Methanobacterium thermoautotrophicum* (42).

The remainder of the spectrum has features that are possibly due to superimposed signals, evidenced by a shoulder at $g = 2.15$ and 2.00, plus a small signal at $g = 1.94$, which is possibly due to a ferredoxin component. The free radical signal at $g = 2.00$ is possibly due to a flavin or quinone electron carrier. A flavin peak is observed in difference spectra of *N. patriciarum* whole cells and hydrogenosomes (N. Yarlett and C. G. Orpin, unpublished results), and flavin acts as an acceptor for hydrogenase from this organism (7).

B. Sensitivity to Oxygen

The rate of hydrogen production by *N. patriciarum* under anaerobic conditions is about 8.1 nmol min^{-1}. As with other hydrogenases from anaerobic organisms, hydrogen production by *N. patriciarum* is inhibited by oxygen, with an apparent K_i (50% inhibition of hydrogen production) of 1.4 μM O_2 (43). Exposure to partial pressures of O_2 that caused complete inhibition of hydrogen production resulted in 30% reduction of hydrogen formation upon return to anaerobic conditions. Assuming that all of the reducing equivalents normally used in the hydrogenase reaction are diverted to reduce oxygen, then the ratio of the difference in the amount of hydrogen produced before

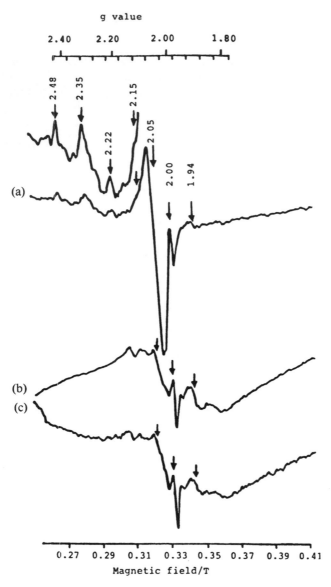

Figure 3 Electron spin resonance spectra of *Neocallimastix patriciarum* hydrogenosomes. **(a)** Air-oxidized, **(b)** incubated with 10 mM pyruvate and 1 mM coenzyme A, **(c)** reduced with 5 mM dithionite. Spectra were recorded at 21 K with the following instrument settings: microwave power 10 mW, frequency 9.19 GHz. Expanded regions are fivefold increased gain. (N. Yarlett, R. Cammack, D. Lloyd, unpublished data.)

and after oxygen exposure to that of oxygen uptake enables a product balance of 2:1 to be calculated, indicating that water is the major product under these conditions. At oxygen tensions above 10 μM in solution, this ratio changes to 3.7:1 (43), a feature previously observed with the rumen protozoan *Dasytricha ruminantium* (44) and which is assumed to be the result of inactivation of hydrogenase either directly by oxygen or by products of oxygen reduction, e.g., peroxide, superoxide, etc. In common with other hydrogenosome-containing eucaryotes, *N. patriciarum* exhibits an apparent oxygen affinity of 4.0 μM, which is comparable to that of aerobic protozoa possessing well-developed mitochondria (45). Hence *Neocallimastix* is able to cope with the presence of low levels of oxygen, and its oxygen-scavenging systems are capable of rapidly consuming oxygen at concentrations of 4 μM or less (43). The role of oxygen in energy production by the fungus is not clear at this time, although it has recently been shown that both oxygen and carbon dioxide affect hydrogenosomal electron transport by *Trichomonas vaginalis* (46). The stimulation of growth and changes in metabolic end products observed in *T. vaginalis* exposed to traces of oxygen (< 0.25 μM) in the presence of 5 mM CO_2 is proposed to be due to the energy gain from enhanced electron flow through the hydrogenosome (46). Low levels of oxygen may have a similar effect in *Neocallimastix*, in which growth under varying carbon dioxide levels has been shown to affect the activity of hydrogenosomal enzymes (43).

Exposure to liquid-phase oxygen concentrations above 10 μM resulted in irreversible inactivation of respiration (43). The concentration of oxygen in rumen contents fluctuates depending on nutrient supply, and the highest level measured in 16 animals was 1.6 μM (47), well below the threshold of oxygen inactivation of respiration. Other species of Neocallimasticaceae are also able to tolerate oxygen for quite long periods without adverse effects (48). Exposure of *N. patriciarum* to air, however, results in the formation of both superoxide and peroxide (43), and *N. patriciarum* was demonstrated to possess superoxide dismutase (SOD) activity but not catalase (43). This is in common with other chytridiomycetes, which also lack catalase (49), and is in line with the theory of oxygen toxicity postulated by McCord et al. (50). In this respect the chytridiomycetes have further similarities to the aerotolerant hydrogenosome-containing trichomonads, some species of which also have no detectable catalase (51). Trichomonads possess a cyanide-insensitive iron-containing SOD, which is procaryotic in nature (52,53). Oomycetes also have a cyanide-insensitive form; in contrast to the chytridiomycetes which possess a cyanide-sensitive form (49). The intracellular location and cyanide-sensitivity of SOD from *N. patriciarum* are not known. The activity of SOD (4.8 units per mg of protein [43]) in *N. patriciarum* is of the same order as that found in other fungi and protozoa (49,52).

IV. FORMATION OF ACETATE

Oxidation of pyruvate by anaerobes necessitates transfer of electrons to acceptors with a more negative redox potential than NAD, in most cases to ferredoxin or flavodoxin. Thus the high reducing power of pyruvate can be used for reactions requiring stronger reductants than NADH, for example, the synthesis of other 2-oxoacids, the reduction of CO_2 to formate, or the excretion of redox equivalents in the form of H_2 (10). Hydrogenosomal acetate formation originates from acetyl-CoA, which in turn is formed from the metabolism of pyruvate by pyruvate:ferredoxin oxidoreductase (7) (Fig. 1). Pyruvate:ferredoxin oxidoreductases have been purified from different organisms and have been shown to have a molecular weight of 200–300 kD, making them much smaller than the pyruvate dehydrogenase complex from aerobic cells. The enzymes from *Clostridium acidi urici* and *Halobacterium halobium* are thiamine diphosphate–containing iron–sulfur proteins (54, 55). Two major types of ferredoxin have been characterized according to molecular and phylogenetic properties. The 4Fe-4S cluster ferredoxins are present in all branches of the Archaeobacteria and are common among the photosynthetic and fermentative anaerobes; hence, this appears to be an ancient phenotype. The 2Fe-2S ferredoxins, however, are common to the Cyanobacteria and function in photosynthetic electron transport. These 2Fe-2S ferredoxins can also act as acceptors for pyruvate oxidation, which is possibly their primordial function (10). The hydrogenosome-containing protozoa of the trichomonad group contain a binuclear Fe-S cluster that exhibits axial electron paramagnetic resonance (EPR) symmetry similar to the hydroxylase-type ferredoxins and lacks the rhombic distortion of chloroplast-type ferredoxins (56). The trichomonad ferredoxin is localized in the hydrogenosome and interacts with pyruvate:ferredoxin oxidoreductase and hydrogenase, supporting the idea that hydrogen production in these eucaryotes is similar to the process occurring in anaerobic bacteria, such as clostridial species. In the clostridia, however, a bitetranuclear ferredoxin serves as electron carrier between pyrivate:ferredoxin oxidoreductase and hydrogenase (57). Although a similar type of ferredoxin has been demonstrated in another anaerobic protozoon, *Entamoeba histolytica*, this organism lacks hydrogenosomes and hydrogenase activity (58). The type of Fe-S cluster in the ferredoxin from *N. patriciarum* has not been determined, nor has it been for the anaerobic rumen ciliates, which also produce hydrogen and contain hydrogenosomes (9); this information would, however, significantly increase our understanding of the phylogenetic evolution of hydrogenosomes.

    The mechanism of acetate production from acetyl-CoA by the rumen chytridiomycetes remains to be identified. It is likely that, in common with other hydrogenosome-containing microorganisms, the end product of the fungal hydrogenosome metabolism is also acetate. The mechanism of acetate formation by the parasitic trichomonads and the symbiotic rumen ciliates differs in the mechanism by which energy is conserved in the form of ATP. In trichomonads a succinate thiokinase functions in combination with an acetyl-CoA:succinate transferase to form acetate and ATP, releasing the bound coenzyme A and thus enabling further formation of acetyl-CoA (59) (Fig. 4a). In some hydrogenosome-containing ciliates the combined action of phosphotransacetylase and acetate kinase results in the formation of acetate and ATP (Fig. 4b) (60). Other possible pathways involve acetate thio-

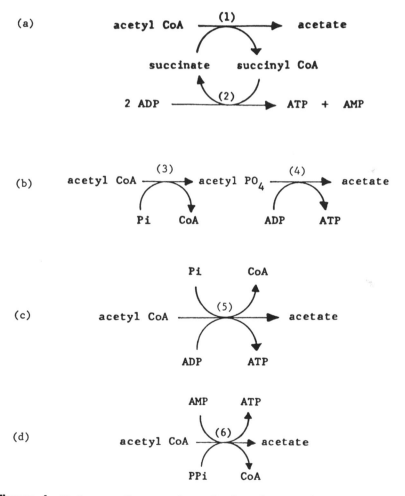

Figure 4 Pathways of acetate formation by microorganisms. (1) acetyl-CoA: succinate transferase; (2) succinate thiokinase; (3) phosphotransacetylase; (4) acetate kinase; (5) acetate thiokinase; (6) acetyl-CoA synthetase.

kinase, which converts acetyl-CoA to acetate, with the concomitant phosphorylation of adenosine diphosphate or guanosine diphosphate (Fig. 4c); and it has also been proposed that some helminths contain an acetyl-CoA synthetase operating in the reverse direction (61) (Fig. 4d).

V. FORMATE AND LACTATE PRODUCTION

For strict and facultative anaerobes the production of formate avoids pyruvate oxidation altogether and hence prevents excessive reduction of the pyridine nucleotide pool. Formate is a major fermentation end product of many of the rumen fungi that have been cultured (12,13,62). In *Neocallimastix* spp. production of formate varies according to the strain, and in several published determinations the amount of formate formed is quantitatively similar to that of acetate (12,13,62), where stoichiometry of end-product formation is 2 formate:2 acetate:2 lactate:1 ethanol (13). Formate has also been detected as an end product in *Sphaeromonas* spp. and *Piromonas* spp. (62). *N. patriciarum* and *N. frontalis* EB188, however, do not appear to produce significant amounts of formate and do not have detectable levels of formate dehydrogenase or formate hydrogen lyase (7,15,16).

In species producing appreciable amounts of formate, pyruvate presumably is converted to formate and acetyl-CoA by the action of pyruvate:formate lyase. The acetyl-CoA may further lead to the formation of acetate as end product, and this mechanism may explain the observation that exogenous formate inhibits growth and acetate production by *Neocallimastix* sp. R1 (13,63).

Lactate production is high in all rumen fungi that have thus far been studied, usually being in the same range as that of acetate. Lactate dehydrogenase is present in those fungi that have been examined (7,16) and functions to recycle reduced pyridine nucleotides. In trichomonads lactate production is in redox balance, since the NAD^+ reduced by glyceraldehyde–phosphate dehydrogenase is reoxidized by lactate dehydrogenase, making lactate formation independent of the production of other end products (21). Unlike the trichomonads, however, the proportion of lactate production by the Neocallimasticaceae appears to be remarkably constant (from 47.6 to 67 mmol/100 mol hexose in five different isolates) (12,13,62).

VI. ALCOHOL AND POLYALCOHOL PRODUCTION

Apart from H_2, acetate, and lactate, the anaerobic fungi produce a variety of metabolic end products. These include ethanol and polyalcohols (glycerol,

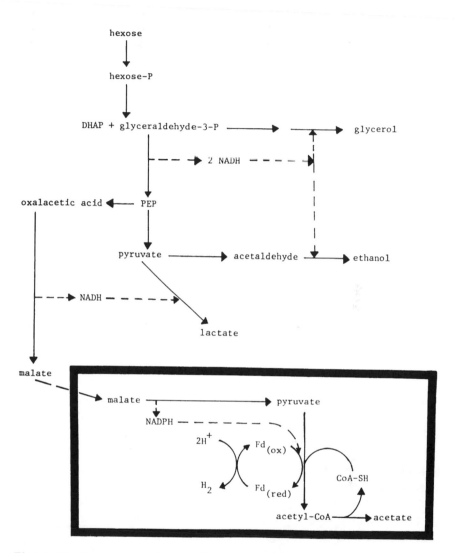

Figure 5 Fermentation redox balance in rumen phycomycetes. The actions of decarboxylating malate dehydrogenase and NADPH:ferredoxin oxidoreductase within the hydrogenosome (boxed area) are self-balancing. Cytoplasmic redox balance is maintained via the production of glycerol, ethanol, and lactate. Dotted lines indicate the proposed flow of redox equivalents. DHAP = dihydroxyacetone phosphate.

erythritol, arabitol, xylitol and ribitol [12,14,64]). However, the metabolic basis for the formation of these products and the fate of reducing power during their synthesis is poorly understood. It has long been known that glycerol is an important by-product of anaerobic alcoholic fermentation in yeasts (65). Formation of both ethanol and glycerol is essentially required to maintain the redox balance (Fig. 5). The formation of ethanol ensures reoxidation of reduced pyridine nucleotides formed in the oxidation of glyceraldehyde-3-phosphate. Glycerol formation serves to maintain the redox balance of cytosolic reactions in generating reduced pyridine nucleotides (Fig. 5). The energetically expensive formation of glycerol may thus be considered to represent a redox value. Both *N. patriciarum* and *Piromonas communis* have recently been shown to produce glycerol as the only polyalcohol fermentation end product (64). Quantitative data on hexose fermentation by *Neocallimastix* spp. (12,14,62) do not show a complete redox balance. Recently, trichomonads were shown to produce glycerol as an additional fermentation end product at a proportion that completes the redox balance of this organism (21). The amount of glycerol produced by the rumen fungi has not been determined, but the published fermentation product balance of various *Neocallimastix* spp. (12,14,62) indicates that an equimolar concentration of ethanol and glycerol would satisfy the redox balance. The enzymes responsible for glycerol production in *Neocallimastix* spp. have not been determined, but most species producing glycerol contain glycerol phosphate dehydrogenase and glycerol-3-phosphatase, which catalyzes the formation of glycerol via glycerol-3-phosphate (66,67).

VII. CONCLUSIONS

Anaerobic microorganisms overcome the problems of maintaining a redox balance in different ways. The rumen chytridiomycetes have many similarities to the hydrogenosome-containing anaerobic protozoa with respect to fermentation pathways. However, many of the enzymes involved in the synthesis of some of the end products, and the mechanisms controlling glycolysis and end-product formation, have not been examined in these anaerobic fungi. In protozoa lacking functional mitochondria, and therefore dependent on a fermentation-based metabolism for ATP production, the key glycolytic enzymes are differentially controlled as compared with protozoa containing functional mitochondria (68). Many intriguing questions concerning the localization and characterization of fermentation pathways in the rumen chytridiomycetes remain to be answered. Studies addressing these questions would provide interesting comparative data concerning the phylogenetic origins of this diverse group.

REFERENCES

1. Orpin CG. The occurrence of chitin in the cell walls of the rumen organisms *Neocallimastix frontalis, Piromonas communis* and *Sphaeromonas communis.* J Gen Microbiol 1977; 99:215-218.
2. Orpin CG. Studies on the rumen flagellate *Neocallimastix frontalis.* J Gen Microbiol 1975; 91:249-262.
3. Orpin CG, Mathieson SD, Greenwood Y, Blix AS. Seasonal changes in the ruminal microflora of the high-arctic Svalbard reindeer (*Rangifer tarandus platyrhynchus*). Appl Environ Microbiol 1986; 50:144-151.
4. Orpin CG. Isolation of cellulolytic phycomycete fungus from the caecum of the horse. J Gen Microbiol 1981; 123:287-296.
5. Orpin CG. Nutrition and biochemistry of anaerobic Chytridiomycetes. BioSystems 1988; 21:365-370.
6. Bauchop T. The gut anaerobic fungi: colonizers of dietary fibre. In: Wallace G, Bell L, eds. Fibre in Human and Animal Nutrition. Wellington, New Zealand: Royal Society of New Zealand 1983:143-148.
7. Yarlett N, Orpin CG, Munn EA, et al. Hydrogenosomes in the rumen fungus *Neocallimastix patriciarum.* Biochem J 1986; 236:729-739.
8. Lindmark DG, Muller M. Hydrogenosome, a cytoplasmic organelle of the anaerobic flagellate *Tritrichomonas foetus,* and its role in pyruvate metabolism. J Biol Chem 1973; 248:7724-7728.
9. Yarlett N, Hann AC, Lloyd D, Williams AG. Hydrogenosomes in the rumen protozoon *Dasytricha ruminantium.* Biochem J 1981; 200:365-372.
10. Kerscher L, Oesterhelt D. Pyruvate:ferredoxin oxidoreductase—new findings on an ancient enzyme. Trends Biochem Sci 1982; 7:371-374.
11. Gottschalk G. Bacterial metabolism. 2nd edition. New York: Springer-Verlag, 1986:39-65.
12. Bauchop T, Mountfort DO. Cellulose fermentation by a rumen anaerobic fungus in both the absence and the presence of rumen methanogens. Appl Environ Microbiol 1981; 42:1103-1110.
13. Lowe SE, Theodorou MK, Trinci APJ. Growth and fermentation of an anaerobic rumen fungus on various carbon sources and effect of temperature on development. Appl Environ Microbiol 1987; 53:1210-1215.
14. Orpin CG. Carbohydrate fermentation in a defined medium by the rumen phycomycete *Neocallimastix frontalis.* Proc Soc Gen Microbiol 1978; 7:31-32.
15. Orpin CG, Munn EA. *Neocallimastix patriciarum* sp. nov., a new member of the Neocallimasticaceae inhabiting the rumen of sheep. Trans Br Mycol Soc 1986; 86:178-181.
16. O'Fallon JV, Wright RW Jr, Calza RE. Glucose metabolic pathways in the anaerobic rumen fungus *Neocallimastix frontalis* EB188. Biochem J 1991; 274:595-599.
17. Reymond P, Geourjon C, Roux B, et al. Sequence of the phosphoenolpyruvate carboxylase-encoding cDNA from the rumen anerobic fungus *Neoxallimastix frontalis*: comparison of the amino acid sequence with animals and yeast. Gene 1991; 110:57-63.

18. Lo H-S, Reeves R. Pyruvate-to-ethanol pathway in *Entamoeba histolytica*. Biochem J 1978; 171:225–230.

19. Lindmark DG. Energy metabolism of the anerobic protozoon *Giardia lamblia*. Mol Biochem Parasitol 1980; 1:1–12.

20. Czerkasovova A, Cerkasov J, Kulda J. Metabolic differences between metronidazole resistant and susceptible strains of *Tritrichomonas foetus*. Mol Biochem Parasitol 1984; 11:105–118.

21. Steinbuchel A, Müller M. Glycerol, a metabolic end product of *Trichomonas vaginalis* and *Tritrichomonas foetus*. Mol Biochem Parasitol 1986; 20:45–55.

22. Munn EA, Orpin CG, Hall FJ. Ultrastructural studies of the free zoospore of the rumen phycomycete, *Neocallimastix frontalis*. J Gen Microbiol 1981; 125: 311–323.

23. Heath IB, Bauchop T, Skipp RA. Assignment of the rumen anaerobe *Neocallimastix frontalis* to the Spizellomycetales (Chytridiomycetes) on the basis of its polyflagellate zoospore ultrastructure. Can J Bot 1983; 61:295–307.

24. Webb J, Theodorou MK. A rumen anaerobic fungus of the genus Neocallimastix: ultrastructure of the polyflagellate zoospore and young thallus. BioSystems 1988; 21:393–401.

25. Heath IB. Ultrastructure of fresh water phycomycetes. In: Jones EBJ, ed. Recent Advances in Aquatic Mycology. London: Paul Elek Press, 1976:603–650.

26. Brugerolle G. Caractérisation ultrastructurale et cytochimique de deux types de granules cytoplasmiques chez les Trichomonas. Protistologica 1972; 8:353–363.

27. Yarlett N, Hann AC, Lloyd D, Williams AG. Hydrogenosomes in a mixed isolate of *Isotricha prostoma* and *Isotricha intestinalis* from ovine rumen contents. Comp Biochem Physiol 1983; 74B:357–364.

28. Gray CT, Gest H. Biological formation of molecular hydrogen. Science 1965; 148:186–192.

29. Bauchop T. Mechanism of hydrogen formation in *Trichomonas foetus*. J Gen Microbiol 1971; 68:27–33.

30. Mountfort DO, Asher RA, Bauchop T. Fermentation of cellulose to methane and carbon dioxide by a rumen anaerobic fungus in a triculture with *Methanobrevibacter* sp. strain RA1 and *Methanosarcina barkeri*. Appl Environ Microbiol 1982; 44:128–134.

31. Marvin-Sikkema FD, Richardson AJ, Stewart CS, Gottschal JC, Prins RA. Influence of hydrogen-consuming bacteria on cellulose degradation by anaerobic fungi. Appl Environ Microbiol 1990; 56:3793–3797.

32. Mountfort DO, Asher RA. Production and regulation of cellulase by two strains of the rumen anaerobic fungus *Neocallimastix frontalis*. Appl Environ Microbiol 1985; 49:2314–2322.

33. Wolin MJ. Metabolic interactions among intestinal microorganisms. Am J Clin Nutr 1974; 27:1320–1328.

34. Stredas T, Antanaitis BC, Krasna AI. Characterization and stability of hydrogenase from Chromatium. Biochim Biophys Acta 1980; 616:1–9.

35. Adams MWW, Hall DO. Purification of membrane-bound hydrogenase of *Escherichia coli*. Biochem J 1979; 183:11–22.

36. Schoenmaker GS, Oltman LF, Stouthamer AH. Purification and properties of the membrane-bound hydrogenase from *Proteus mirabilis*. Biochim Biophys Acta 1979; 567:511–521.

37. Mayhew SG, O'Connor ME. Structure and mechanism of bacterial hydrogenase. Trends Biochem Sci 1982; 7:18–21.

38. Fauque G, Teixeira M, Moura I, et al. Purification, characterization and redox properties of hydrogenase from *Methanosarcina barkeri* (DSM800). Eur J Biochem 1984; 142:21–28.

39. Schneider K, Patil DS, Cammack R. ESR properties of membrane-bound hydrogenase from aerobic hydrogen bacteria. Biochim Biophys Acta 1983; 748:353–361.

40. Adams MWW, Mortenson LE, Chen JS. Hydrogenase. Biochim Biophys Acta 1980; 594:105–176.

41. Cammack R, Patil D, Aguirre R, Hatchikian EC. Redox properties of the ESR-detectable nickel in hydrogenase from *Desulfovibrio gigas*. FEBS Lett 1982; 142:289–292.

42. Albracht SPJ, Graf E-G, Thauer RK. The EPR properties of nickel in hydrogenase from *Methanobacterium thermoautotrophicum*. FEBS Lett 1982; 140:311–313.

43. Yarlett N, Rowlands C, Yarlett NC, et al. Respiration of the hydrogenosome-containing fungus *Neocallimastix patriciarum*. Arch Microbiol 1987; 148:25–28.

44. Yarlett N, Scott RI, Williams AG, Lloyd D. A note on the effects of oxygen on hydrogen production by the rumen protozoon *Dasytricha ruminantium* Schuberg. J Appl Bacteriol 1983; 55:359–361.

45. Lloyd D, Williams J, Yarlett N, Williams AG. Oxygen affinities of the hydrogenosome-containing protozoa *Tritrichomonas foetus* and *Dasytricha ruminantium*, and two aerobic protozoa determined by bacterial bioluminescence. J Gen Microbiol 1982; 128:1019–1022.

46. Paget TA, Lloyd D. *Trichomonas vaginalis* requires traces of oxygen and high concentrations of carbon dioxide for optimal growth. Mol Biochem Parasitol 1990; 41:65–72.

47. Scott RI, Yarlett N, Hillman K, et al. The presence of oxygen in rumen liquor and its effects on methanogenesis. J Appl Bacteriol 1983; 55:143–149.

48. Trinci APJ, Lowe SE, Milne A, Theodorou MK. Growth and survival of rumen fungi. BioSystems 1988; 21:357–363.

49. Natvig DO. Comparative biochemistry of oxygen toxicity in lactic acid–forming aquatic fungi. Arch Microbiol 1982; 132:107–114.

50. McCord JM, Keele BB, Fridovich I. An enzyme-based theory of obligate anaerobiosis: the physiological function of superoxide dismutase. Proc Natl Acad Sci USA 1971; 68:1024–1027.

51. Lindmark DG, Müller M. Biochemical cytology of trichomonad flagellates. II. Subcellular distribution of oxidoreductases and hydrolases in *Monocercomonas* sp. J Protozool 1974; 21:374–378.

52. Lindmark DG, Müller M. Superoxide dismutase in the anaerobic flagellate *Tritrichomonas foetus* and *Monocercomonas* sp. J Biol Chem 1974; 249:4634–4637.

53. Kitchener KR, Meshnick SR, Fairfield AS, Wang CC. An iron containing super-oxide dismutase in *Trichomonas foetus*. Mol Biochem Parasitol 1984; 12:95–99.
54. Uyeda K, Rabinowitz JC. Pyruvate–ferredoxin oxidoreductase III. J Biol Chem 1971; 246:3111–3119.
55. Kerscher L, Oesterhelt D. Purification and properties of two 2-oxoacid:ferre-doxin oxidoreductases from *Halobacterium halobium*. Eur J Biochem 1981; 116:587–594.
56. Gorrel TE, Yarlett N, Müller M. Isolation and characterization of *Trichomonas vaginalis* ferredoxin. Carlsberg Res Commun 1985; 49:259–268.
57. Yasunobu KT, Tanaka M. The types, distribution in nature, structure-function, and evolutionary data of the iron sulfur proteins. In: Lovenberg W, ed. Iron Sulfur Proteins. Vol. II, Moelcular Properties. New York: Academic Press. 1973:127–130.
58. Reeves RE, Guthrie JD, Lobelle-Rich R. *Entamoeba histolytica*: isolation of ferredoxin. Exp Parasitol 1980; 49:83–88.
59. Lindmark DG. Acetate production by *Tritrichomonas foetus*. In: Vanden Bossche H, ed. Biochemistry of Parasites and Host Parasite Relationships. Amsterdam: Elsevier/North Holland. 1976:15–21.
60. Yarlett N, Lloyd D, Williams AG. Respiration of the rumen ciliate *Dasytricha ruminantium* Schuberg. Biochem J 1982; 206:259–2667.
61. Saz HJ. Comparative energy metabolism of some parasitic helminths. J Parasitol 1970; 56:634–642.
62. Phillips MW, Gordon GL. Sugar and polysaccharide fermentation by rumen anaerobic fungi from Australia, Britain and New Zealand. BioSystems 1988; 21:377–383.
63. Theodorou MK, Lowe SE, Trinci APJ. The fermentative characteristics of anaerobic rumen fungi. BioSystems 1988; 21:371–376.
64. Pfyffer GE, Boraschi-Gaia C, Weber B, et al. A further report on the occur-rence of acyclic sugar alcohols in fungi. Submitted.
65. VanDijken JP, Scheffers A. Redox balances in the metabolism of sugars by yeasts. FEMS Microbiol Rev 1986; 32:199–224.
66. Gancedo C, Gancedo JM, Sols A. Glycerol metabolism in yeasts. Pathways of utilization and production. Eur J Biochem 1968; 5:165–172.
67. Van Schaftingen E, VanLaere AJ. Glycerol formation after breaking of dor-mancy of *Phycomyces blakesleeanus* spores. Role of an interconvertible glycerol-3-phosphatase. Eur J Biochem 1985; 148:399–404.
68. Mertens E, Van Schaftingen E, Müller M. Presence of a fructose-2,6-biphos-phate–insensitive pyrophosphate:fructose-6-phosphate phosphotransferase in the anaerobic protozoa *Tritrichomonas foetus, Trichomonas vaginalis* and *Iso-tricha prostoma*. Mol Biochem Parasitol 1989; 37:183–190.

5

Regulatory Constraints in the Degradation and Fermentation of Carbohydrate by Anaerobic Fungi

DOUGLAS O. MOUNTFORT

Cawthron Institute
Nelson, New Zealand

I. INTRODUCTION

Until the late 1970s it was believed that metabolic transformations in anaerobic ecosystems such as the rumen were carried out by bacteria and protozoa. Fungi were essentially considered as aerobic and were not expected to exist in anaerobic ecosystems. It was therefore not surprising that the first flagellates to be described in the rumen (1,2) were regarded as protozoa and received little attention until the studies of Orpin, who demonstrated that the flagellates were the motile forms of anaerobic fungi. It was proposed that the fungi possessed a life cycle alternating between a motile flagellate form or zoospore and a nonmotile vegetative reproductive form or thallus consisting of a rhizoid-bearing sporangium.

Anaerobic fungi have been isolated mainly from the rumen of cattle and sheep (3–16), but there are also isolates from the cecum of the horse (17,18, 19), and the organisms have been shown to be present in the guts of feral ruminants and marsupials (20). Most of the anaerobic fungi that have been described fall within four genera. These are *Neocallimastix* (12), *Sphaeromonas* (13) (= *Caecomyces* [17]), *Piromonas* (14) (= *Piromyces* [3]), and *Orpinomyces* (3). Members of these genera have all been shown to possess chitin in their cells walls (21,22).

In vivo studies have established that anaerobic fungi extensively colonize plant material suspended in the rumen in nylon bags, and that the principal sites of invasion are on damaged tissues (14,23,24,25). Because of the widespread colonization of plant material by fungi it has been suggested that they have a role in fiber digestion and are the initial colonizers in lignocellulose digestion (23-26). This view was supported by the studies of Akin et al. (27), who observed that anaerobic fungi were prevalent in the rumen of sheep fed on the forage *Digitaria pentzii* that had been fertilized with sulfur. Nylon-bag trials showed extensive colonization by fungi of lignified cells of leaf blade sclerenchyma tissue, and it was likewise suggested that these organisms had a role in lignocellulose digestion. The significance of anaerobic fungi in fiber digestion was demonstrated in in vitro digestion trials showing that in the absence of an actively growing bacterial population, almost 60% of forage material could be removed by these organisms.

II. FERMENTATION OF CARBOHYDRATE BY ANAEROBIC FUNGI AND REGULATORY MECHANISMS

Studies on fermentations by anaerobic fungi became possible after they were isolated in pure culture. Some of the first studies on the action of anaerobic fungi on pure substrates were carried out by Bauchop (23) and Orpin and Letcher (28). These workers demonstrated the colonization of Whatman no. 1 filter paper strips by fungi, and this was accompanied by extensive pitting of the paper. Cellulosic digestion was quantitated in time-course experiments showing that growth of *Neocallimastix frontalis* on filter paper strips was accompanied by a decrease in cellulose: at full growth between 50% and 70% of the initial cellulose could be digested. The first comprehensive study of a fermentation by an anaerobic fungus was carried out by Bauchop and Mountfort (4), who showed that *N. frontalis* fermented cellulose to give six major products. Moles products as a percent of moles cellulosic hexose fermented were acetate, 72.7; carbon dioxide, 37.6; formate, 83.1; ethanol, 37.4; lactate, 67.0; and H_2, 35.3. Similar fermentation profiles for glucose and cellulose have since been described by other workers for a number of rumen fungal isolates (5,16,29,30,31). Hydrogen production appears to be a feature of nearly all anaerobic fungal fermentations, and this property allows these organisms to be cocultured with H_2-utilizing organisms (4). In the cocultures there is a shift in the fermentation pattern toward the formation of more oxidized products such as acetate (Table 1), as has also been observed with rumen bacterial H_2producers in methanogenic coculture (32,33,34). It is thought that reducing equivalents otherwise used in the production of electron-sink products such as lactate and ethanol in fungal monoculture are diverted via H_2 to toward the formation of reduced products such as methane

Table 1   Fermentation of Cellulose by *Neocallimastix frontalis* in the Absence and Presence of *Methanobrevibacter ruminantium*[a]

Product	mol product/100mol hexose units[b]	
	N. frontalis	plus *M. ruminantium*
Acetate	72.7	134.7
Lactate	67.0	2.9
Ethanol	37.4	19.0
Methane	0.0	58.7
Carbon Dioxide	37.6	88.7
Hydrogen	35.3	<0.05
Formate	83.1	1.0

[a]Determined at the completion of fermentation, at which time the percentage of cellulose degraded in the monoculture was 53%, and in the coculture, 82%.
[b]Data after Bauchop and Mountfort (4).

in the cocultures via a fungal pyridine nucleotide–linked hydrogenase. The hydrogenase-mediated electron transfer becomes favorable through a reduction in the H_2 partial pressure by the H_2 utilizer (35). The site of H_2 production in the anaerobic fungi is on intracellular organelles called "hydrogenosomes" (36,37), which are different from mitochondria but similar to the H_2-evolving organelles of certain anaerobic protozoa (38,39). In the hydrogenosomal fraction obtained from *Neocallimastix patriciarum*, hydrogenase, pyruvate:ferredoxin oxidoreductase, NADPH:ferredoxin oxidoreductase, and malic enzyme have been found to be enriched, and Yarlett et al. (37) proposed that these enzymes were involved in H_2 formation according to the scheme shown in Fig. 1. A similar scheme employing malate taken up by hydrogenosomes has been proposed by O'Fallon et al. (30) for hydrogen production in *N. frontalis* strain EB 188. However, the origin of oxaloacetate, the immediate precursor of malate, is believed to be from pyruvate rather than from phosphoenolpyruvate (PEP), on the basis of the demonstration of pyruvate carboxylase activity and the absence of PEP carboxykinase in enzyme preparations from this organism. Figure 1 also shows pathways for the formation of the other fermentation end products by *N. patriciarum*, on the basis of the demonstration of the appropriate enzymic activities in either the hydrogenosomal or the cytosolic fractions of this organism. Reactions leading to the formation of formate are not included since unlike other anaerobic fungi, *N. patriciarum* does not produce this as a fermentation end product (37). Absence of glucose-6-phosphate dehydrogenase and the demonstration of all the key glycolytic enzymes in this organism suggest that hexose

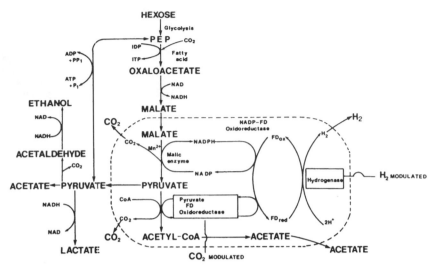

Figure 1 Pathways for the production of hydrogen and other fermentation end products from hexose fermentation by *N. patriciarum*, and sites of metabolic control. Enclosed area refers to reactions occurring within the hydrogenosome. (Adapted from Ref. 37.)

fermentation occurs by the Emden-Meyerhof-Pathway. This pathway also appears to be the major route for hexose fermentation in *N. frontalis* EB 188 as established by radiolabeling procedures (30). As yet there is little or no information on the fermentation enzymes in other genera of anaerobic fungi, but on the basis of their fermentation patterns it is likely that the pathways in these organisms are similar to those of *Neocallimastix*. Absence of key enzymes of the tricarboxylic acid in *Neocallimastix* (30,37) suggests that degradation occurs wholly by fermentative mechanisms.

Utilization of H_2 by methanogens and other H_2 utilizers in fungal cocultures, apart from shifting the fermentation pattern, appears to have a second effect, which is to increase the rate and extent of carbohydrate utilization. This phenomenon was first observed by Bauchop and Mountfort (4), who showed that a coculture of *N. frontalis* with *Methanobrevibacter* degraded cellulose at a faster rate than a culture with the fungus alone. Furthermore, at the end of growth 82% of the initial cellulose had been utilized by the coculture and 53% by the monoculture. When the fungus was grown in triculture with H_2-utilizing and acetate-utilizing methanogens, complete cellulose degradation could be obtained (40). This effect has since been demonstrated with different carbohydrate substrates and with different fungus–H_2 utilizer combinations (41) and is dealt with comprehensively in Chapter 7. One possible explanation for increased carbohydrate degradation by anaerobic fungi in the presence of H_2-utilizing organisms is that electron-sink products otherwise inhibitory in monocultures are removed. Mountfort et al. (40) suggested that lactate and ethanol could be inhibitors of cellulose utilization by *N. frontalis*, as decreased production of these by the altered fermentation in the presence of methanogens was linked to increased substrate utilization. Evidence that these products are inhibitors has been obtained from studies on cellulolysis and xylanolysis by fungi in the presence and absence of lactate-utilizing rumen organisms, in which it was demonstrated that degradation was increased in the cocultures (42, Chapter 7), and from studies on the growth of fungal monocultures in product-supplemented media (43). At present, it is not yet clear whether lactate and ethanol exert their effect on the polysaccharidases or on the fermentation pathway, or on both. Acetate does not appear to exert a negative effect on fermentations, since increased substrate utilization in fungus–H_2-utilizing methanogen cocultures occurs with increased acetate production (4).

Apart from the influence of H_2 on fermentations in anaerobic fungi, little is known about other regulatory constraints. In *N. patriciarum* CO_2 has been found to suppress pyruvate:ferredoxin oxidoreductase (37), which is a key enzyme of H_2 and acetate formation. Synthesis of the enzyme has been found to recommence under CO_2-free conditions. Under conditions of elevated CO_2 such as those found in the rumen, acetate production would be expected to occur via pyruvate in a ferredoxin–independent pathway, and reducing power for hydrogen production would be provided via NAD (NADP):ferredoxin oxidoreductase. Carbon dioxide has also been found to have a repressive effect on PEP carboxykinase synthesis in *N. patriciarum* (37), and the relative roles of this enzymes and other enzymes (possibly pyruvate carboxylase and PEP carboxylase) in the production of oxaloacetic acid may be dependent on the levels of CO_2. Pyruvate carboxylase and PEP carboxykinase are also known to be stimulated by acetyl-CoA and fatty acids in other microorganisms, but these modifiers have a negative effect on pyruvate kinase. It is therefore not surprising that in cell extracts of *N. patriciarum* very low levels of pyruvate kinase occur. Clearly, many of the regulatory aspects of carbohydrate fermentation in anaerobic fungi still require to be elucidated. The targets of end-product action on fermentation pathways that have been defined are indicated in Fig. 1 and are summarized in Table 2.

III. ENZYMES OF ANAEROBIC FUNGI INVOLVED IN POLYSACCHARIDE DEGRADATION, AND REGULATORY MECHANISMS

A. Enzymes That Degrade Cellulose

The finding by Bauchop (23) and Orpin and Letcher (28) that cellulose in the form of Whatman no. 1 filter paper could be extensively colonized by an-

Table 2 Effect of Metabolites in Regulating Carbohydrate Utilization and Fermentation in Anaerobic Fungi

Metabolite	Enzyme or pathway affected	Nature of action	Effective concentration	Reference
CO_2	Pyruvate:ferredoxin oxidoreductase	Repressor	100.0 kPa	37
H_2	Pyridine nucleotide–linked hydrogenase	Inhibition	0.1 kPa	35
Lactate	Unknown	End-product inhibitor	~20.0mM	42,43
Glucose	Cellulase	Repressor	≥22.0 mM	44–47
Xylose	Xylanase	Repressor	≥20.0 mM	66,67
Arabinose	Xylanase	Repressor	≥20.0 mM	67
Glucose	α-amylase	Repressor	≥17.0 mM	76
Glucose	Fructose uptake	Catabolite regulation	≥2.5 mM	82
Glucose	Xylose uptake	Catabolite regulation	≥5.0 mM	82

aerobic fungi led to the suggestion that these organisms might possess unique features, perhaps residing in their cellulases, that could facilitate the deep mycelial penetration and rapid colonization of plant material.

For the hydrolysis of cellulose three enzymes are necessary. these are endo-β-1,4-glucanase, exo-β-1,4-glucanase, and β-glucosidase.The first two enzymes act in concert to produce cellobiose and cellodextrins from cellulose, and these are metabolized further by β-glucosidase to produce glucose.

1. Endo-β-1,4-Glucanase and exo-β-1,4-Glucanase

Endo-β-1,4-glucanase (carboxymethylcellulase [CMCase]) and exo-β-1,4-glucanase (avicelase or cellobiohydrolase) activities are known to be present in the vegetative and zoospore stages of the life cycle of *N. frontalis*, *N. patriciarum*, and *Piromonas communis*, and the enzymes are also released into the culture fluid (44–50). In *N. frontalis* CMCase is mainly released into the culture fluid as treatment of rhizoid with various solubilizing agents such as Tween 80 or toluene–ethanol–Triton-X-100, released enzyme with activity accounting for less than 10% of the total activity of the cultures (47). On the other hand, cellobiohydrolase appears to be mainly cell wall–bound, and the activity of this enzyme may be dependent on its association with the cell wall for maintenance of its conformation for attacking crystalline cellulose (49). In the case of *P. communis* and *N. patriciarum*, it appears that a substantial amount of cellulase is also retained by the fungal rhizoid (48).

Optimum production cellulase by anaerobic fungi occurs with the cellulosic growth substrate at about 2.0 mg ml^{-1} in culture media (47,48). When the growth substrate is increased beyond this value, cellulase production declines and glucose may accumulate in the growth medium to as much as 20 mM. Glucose appears to have a regulatory effect on cellulase at elevated concentrations, because its addition to cultures of *N. frontalis* actively producing cellulase from growth on cellulose suppresses enzyme production (47). Similarly, when glucose-grown cultures are switched to cellulose there is an increase in cellulase excretion (44,45,46). Thus, while cellulase production can occur in the presence of low levels of glucose, the enzyme is suppressed by elevated levels of this sugar. The mechanism for glucose action appears to be repression rather than inhibition (see Table 2 for summary of effects), because there is no decline in the activity of cellulase enzyme preparations upon addition of glucose at the same concentrations that suppressed activity in growing cultures (47).

It is thought that the endo- and exo-acting glucanases of anaerobic fungi are produced initially as a large multienzyme complex probably associated with the membrane fraction of the fungal cell wall and subsequently released into the culture fluid. In *N. frontalis* the complex is believed to be between 750 and 1000 kDa in size, although its composition varies depending on the growth conditions (49). The amount of the complex is also influenced by growth conditions. Thus, cultures grown in defined medium are rich in high-molecular-weight complex, whereas those grown in media containing rumen fluid have low levels. Studies with chitinase suggest that one component of the complex, an exoglucanase, is dependent on its association with the cell wall to maintain the conformation required to attack crystalline cellulose. Thus, not all cellulases produced by anaerobic fungi are necessarily truly cell-free.

Seven CMCases and six avicelases have been identified in the culture fluids of *N. frontalis* EB188 (51). The relationship between these cellulases is not clear, and proteolytic action of proteases secreted into the culture medium (52,53) may indicate a higher number of cellulases than were actually produced. The degree of glycosylation of the enzymes may also be important in determining whether they are excreted, and their substrate specificities (54).

The ability of *Neocallimastix* cellulases to retain their activity when not glycosylated has enabled a range of plant cell–hydrolyzing genes to be isolated from the cDNA library prepared from *N. patriciarum* mRNA, and constructed in *Escherichia coli* using the lambda ZAPII vector (55,56,57). Some of the recombinant genes isolated from this library encode cellulase and xylanases containing more than one active site, and some have active sites with multiple substrate specificities (56,57).

Five different cellulase genes have been cloned from *N. patriciarum* (55, 56,57), and they all express proteins in *E. coli* with cellodextrinase and endo-b-1,4-glucanase activities. Three of them (*cel A*, *cel D*, and *cel E*) also have significant cellobiohydrolase activites and release cellobiose from crystalline cellulose, while *cel D* also has xylanase and xylo-oligosaccharase activities

(57). In addition, *cel B* and *cel D* have cellobiosidase activity. *Cel A*, *cel B*, and *cel C* are inducible by the presence of cellulose, while *cel D* is produced constitutively. These findings tend to confirm the earlier suggestions that cellulases in anaerobic fungi are partially constitutive because of the considerable levels of activity obtained when cultures were grown on a variety of noncellulosic sugars and are also partially inducible because of the significantly higher levels of activity obtained after growth on cellulose as compared with noncellulosic substrates such as glucose (44,47).

The constitutive enzyme expressed by *cel D* may be of major importance to the fungus during initial colonization of a plant fragment in the competitive environment of the rumen. During this process, which may occur as soon as 10 min after the host animal has started eating, a free zoospore locates on a freshly ingested plant fragment and subsequently encysts and germinates. A penetration rhizoid grows into the plant tissue and anchors the young fungus to its substratum by penetration of the cell walls. During this period some soluble carbohydrates, particularly glucose, may be present in the rumen liquor at relatively high concentrations. Repression of cellulases by glucose or cellobiose is known to occur in anaerobic fungi (47) at elevated concentrations of these sugars. Thus, if all cellulases were subject to repression there would be a delay in the colonization of plant tissues, and the competitive advantage gained by rapid invasion of the particle by the zoospore would be lost. In addition, the *cel D* enzyme has a wide specificity in being able to attack crystalline and amorphous forms of cellulose in addition to xylans. This ability would seem ideal for the rapid penetration of plant cell polysaccharides by the fungus, removing the wait for the production of the inducible enzymes.

Native cellobiohydrolases produced in cultures of *N. patriciarum* occur in several forms that release cellobiose from crystalline cellulose. In actively growing cultures the molecular weights of isolated cellobiohydrolases range from 46 to 60 kDa, whereas a form with a molecular weight of 60 kDa has been isolated from vegetative growth (58). The relationship between these forms has yet to be determined, but the ultimate formation of low-molecular-weight cellobiohydrolase may be beneficial to the deep penetration of enzyme into the structure of the plant cell wall. The relationship of these cellobiohydrolases to the recombinant cellulases with cellobiohydrolase activity (*cel A*, *cel D*, and *cell E*) are not yet known.

The comparatively low levels of cellobiohydrolase as compared with CMCase obtained by different workers have led authors to consider whether Avicel and Sigmacel are the best substrate for measurements of enzymatic activity. Mountfort and Asher (47) prepared short-chain cello-oligosaccharides and obtained higher levels of activity with these than with Avicel. Their conclusion was that cello-oligosaccharides afforded a better indication of cellobiohydrolase activity. However, it seems more likely that the shorter-chain oligomers simply afforded a measure of cellodextrinase. Cellodextrinase has now been detected with all the cellulase genes expressed from recombinant genes isolated from cDNA libraries prepared from *N. patriciarum* (55–57).

2. β-Glucosidase

Glycosidases are necessary for the complete degradation of plant cell polysaccharides, and β-D-glucosidase is required for the complete degradation of cellulose to glucose. The latter enzyme has been detected in the culture fluids of *N. patriciarum*, *N. frontalis*, and *Piromyces* sp. (59) and in other rumen fungus cultures. Four different glucosidases differentiated by molecular weight occur in *N. frontalis* EB 188 culture fluids (60). Two of these have been purified and characterized, and their properties are summarized in Table 3. These enzymes readily hydrolyze *p*-nitrophenyl β-D-glucopyranoside (pNPG) as well as cellobiose, and activity is inhibited by glucose and glucono-δ-lactone during the hydrolysis of pNPG. One of these enzymes also has a low activity toward cellopentaose (61). A glucosidase from *Piromyces* sp. has also been purified and characterized but has a substantially lower isoelectric point and molecular weight than the enzymes described for *N. frontalis* (62) (Table 3). In addition, the enzyme has a broad substrate range with ability to actively hydrolyze cello-oligosaccharides with a chain length of up to five hexose units.

The finding of broad-action glucosidases both in *N. frontalis* and in *Piromyces* indicates that these enzymes should be seen not as simple cleavers of cellobiose during cellulose degradation but rather as having a wider hydrolytic role overlapping with the cellodextrinases of these organisms.

No information exists yet on the regulatory properties of β-glucosidases in anaerobic fungi, but high activities of these enzymes occur following growth on cellulose and cellobiose (59). The enzymes appear to be inducible, because

Table 3 Properties of Glycosidases in Anaerobic Fungi

Enzyme	Organism	MW (kDa)	IEP	pH optimum	K_m (mM)	Reference
β-glucosidase	*N. frontalis*	125	7.10	5.5–7.0	2.50[a]	60
	N. frontalis	85	6.95	5.5–6.8	0.67[a]	61
	Piromyces sp.	45	4.15	6.0	1.50[a]	62
β-xylosidase	*N. frontalis*	150	4.60	6.4	2.42[b]	69
	N. frontalis	180	4.35	6.5	0.33[c]	70

[a]Determined using *p*-nitrophenyl β-glucopyranoside.
[b]Determined using xylobiose.
[c]Determined using *p*-nitrophenyl β-xylopyranoside.
MW = molecular weight; IEP = isoelectric point.

lower levels of activity are obtained after growth on monosaccharides. One study, however, suggests that at least in the case of *Piromyces* sp. glucose and a variety of other mono- and disaccharides are also effective growth substrates for the production of the enzyme (63). This suggests that in the case of this isolate the enzyme may be constitutive.

B. Hemicellulases

The suggestion that anaerobic fungi produce hemicellulases arose from studies by a number of workers who observed degradation of straw and leaf tissue by these organisms. Mountfort et al. (40) observed that the hemicellulose fraction of barley straw leaf and sisal twine was substantially degraded after growth of *N. frontalis* on these substrates in triculture with H_2- and acetate-utilizing strains of methanogens. Similarly, Theodorou et al. (64) demonstrated that arabinose and xylose, which are neutral sugar components of plant cell-walls, were removed during growth of *Neocallimastix* sp. strain R.1 on autoclaved Italian ryegrass hay. Borneman et al. (5) observed that five isolates of anaerobic fungi could degrade coastal Bermuda grass, with a 70% dry-weight loss in 9 days. The demonstration that crude plant tissue could be degraded by anaerobic fungi, together with the knowledge that hemicellulose often composes a substantial proportion of this tissue, strongly indicated that anaerobic fungi possess the necessary enzymes for the breakdown of hemicellulose.

Three enzymes have been identified in anaerobic fungi that are involved in the breakdown of hemicellulose. These are hemicellulase, which is active toward hemicellulose B (obtained from perennial ryegrass, *Lolium perenne*) (48); xylanase, which hydrolyzes β-1,4-linked glycosidic bonds to form xylo-oligosaccharides from xylan (48,63,65–68); and β-xylosidase, which catalyzes the hydrolysis of xylo-oligosaccharides to xylose (69,70).

1. Xylanase

N. patriciarum and *P. communis* appear to retain most of their xylanolytic activity in the vegetative material, and very little xylanase is released into the culture fluid (48). In contrast, *N. frontalis* appears to release most of its xylanase into the culture fluid, and relatively little enzyme is associated with the vegetative material (67). It is possible that the differences reported in the distribution of xylanase between the vegetative material and culture fluid for the various fungi may be attributable to differences in the conditions used to culture the organisms and in the methods for determining enzyme activity.

Like cellulase, xylanase appears to be partially constitutive in anaerobic fungi, as low levels of the enzyme are present after growth on a variety of soluble sugars (48,66,67). However, the enzymes are partially inducible because of the higher levels of activity obtained after growth on xylan or xylan-containing complex substrates such as wheat bran and wheat straw (48,63, 66,67). Of soluble substrates, xylose, starch, and lactose appear to act as effective inducers in the case of *Piromyces* (48,63) while only xylose is an effective inducer in the case of *N. frontalis* and *N. patriciarum* (48,66,67). The highest levels of xylanase are produced by anaerobic fungi grown on xylan, but in the presence of elevated concentrations of the substrate (>2.5 mg ml^{-1}), production of the enzyme declines and xylose and to a lesser extent, arabinose accumulate in the culture medium (66,67). A regulatory effect of these sugars on xylanase production was evident from experiments showing that their addition at >20 mM to actively xylanolytic cultures of *N. frontalis* growing on xylan in which neither sugar had significantly accumulated (<1 mM) suppressed xylanase production (67) (Table 2). Further support for a regulatory role of xylose was obtained in paired-substrate growth experiments with xylan-xylose in which active production of xylanase occurred only during growth on xylan, after xylose had been preferentially utilized (Fig. 2). Thus, while xylose can be an effective inducer of xylanase at low levels, its presence at elevated concentrations leads to repression of the enzyme.

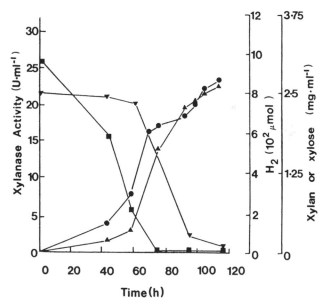

Figure 2 Time-course for xylanase production by *N. frontalis* during growth on a mixture of xylan and xylose at 2.5 and 3.0 mg ml^{-1}, respectively. Symbols: ▲, xylanase activity; ●, hydrogen; ■, xylose; ▼, xylan. (Adapted from Ref. 67.)

Some recent studies have afforded some insight into the nature of xylanases in anaerobic fungi. Teunissen et al. (68) have characterized two extracellular xylanases of *Piromyces* sp. and showed that the enzymes were part of the high-molecular-mass complex present in the culture filtrate. The characteristics of both enzymes were identical, with a molecular mass of 12.5 kDa. The products of the pure enzymes are mainly xylo-oligosaccharides, indicating, that the enzymes are principally endo-acting. The xylanases are inactive against Avicel and carboxymethylcellulose, indicating absence of cellulase activity. Using a different approach, Gilbert et al. (65) have identified a xylanase (xyl A) expressed in *E. coli* from DNA cloned from *N. patriciarum*. The xylanase has a molecular weight of 53 kDa and, like the enzyme characterized for *Piromyces* does not hydrolyze cellulosic substrates. The predicted primary structure of the enzyme comprises an N-terminal signal peptide followed by a 225-amino-acid repeated sequence separated from a tandem 40-residue C-terminal repeat by a threonine/proline linkage sequence. The N-terminal reiterated regions consist of distinct catalytic domains displaying substrate specificities similar to those of the full-length enzyme. The xylanase of *N. patriciarum* is the first reported example of a xylanase that consists of reiterated sequences. Sequence-comparison analysis has revealed significant homology between the xylanase of *N. patriciarum* and bacterial xylanases belonging to the cellulose/xylanase family G. One of these homologous enzymes is derived from the rumen bacterium *Ruminococcus flavefaciens*. This could have been a consequence of horizontal gene transfer between rumen procaryotes and lower eucaryotes.

The two different approaches in elucidating the nature of xylanases of anaerobic fungi (65,68) not only provide valuable insight into the properties of these enzymes but have potential in providing information on their regulatory properties. As with cellulase, the cloning of xylanases may eventually yield information on whether these enzymes are inducible, constitutive, or both. In addition, there appears to be the interesting prospect that at least some fungal xylanases may have been derived from other rumen microflora via gene transfer (65).

2. β-xylosidase

Somewhat less information is available on the nature of β-xylosidases in anaerobic fungi. Purified β-xylosidases have been characterized for *N. frontalis* (69,70), and the properties of these enzymes are summarized in Table 3. The specificity of the enzyme described by Hébraud and Fèvre (70) toward various xylo-oligosaccharides was not determined, but the enzyme described by Garcia Campayo and Wood (69) hydrolyzes a range of xylo-oligosaccharides from xylotriose up to xyloheptaose as well as xylobiose. The xylosidases have little or no activity toward xylan.

No information exists yet on the regulatory properties of xylosidases in anaerobic fungi, but it appears that the highest activities of these enzymes occur following growth on xylan (70). The lower levels of activity obtained after growth on mono- and disaccharides indicate that the enzyme is inducible.

C. Amylases

Starch is a major component of many ruminant diets, and its degradation appears to be related to a number of digestive disorders in ruminants (71). Yet starch-degrading enzymes have been examined in detail in only a few ruminal microorganisms (72–75), and there have been few reports of amylases in ruminal fungi (48,76). Amylolytic enzymes in anaerobic fungi reside in both the vegetative material and in the culture fluid. The proportion of amylase associated with vegetative material or present in the culture fluid varies between different genera and species of fungi, so that in starch-grown *P. communis* cultures most of the amylolytic activity resides in the vegetative material (48), while in the case of *N. frontalis* most of the enzymatic activity is present in the culture fluid (76). Like cellulases and xylanases, amylases in anaerobic fungi are partially constitutive, since activities are present in cultures grown on xylan, cellulose, and a variety of mono- and disaccharides (48). However, since much higher activities are found in cultures grown on starch, it is clear that amylases are also inducible. In *N. frontalis* maltose also appears to be an effective inducer of amylase (76).

Of the amylases examined in anaerobic fungi, only the α-amylase of *N. frontalis* has been studied in some detail (76). The products of this enzyme's action are maltose, maltotriose, maltotetraose, and longer-chain oligosaccharides. The "endo" action of the enzyme was evident from studies showing that the reduction in iodine-staining capacity and release of reducing power by action on amylose was similar to that for commercial α-amylase. In the presence of elevated concentrations of starch, α-amylase production by *N. frontalis* declines, and this is accompanied by an accumulation of glucose in the cultures. Addition of glucose at >17 mM to cultures growing on low levels of starch that do not accumulate the sugar suppresses α-amylase production, and in growth on the paired substrates starch–glucose, active production of α-amylase occurs only during growth on starch after glucose has been preferentially utilized. This suggests that glucose at elevated levels represses α-amylase in *N. frontalis* in a similar way to cellulase (Table 2).

Relatively little study has been carried out on the enzymes that degrade short-chain oligomers and maltose produced from amylase action. Williams and Orpin (48) have demonstrated the activities of α-glucosidases in several anaerobic fungi and shown that the highest activities of these occur after growth on starch or maltose, indicating that the enzymes are inducible.

D. Pectin-Degrading Enzymes

Degradation of pectin is achieved through the action of two enzymes, pectin lyase and polygalacturonase. Until recently anaerobic fungi were considered to have low pectinolytic activity as determined from the production of reducing sugars from pectin. However, Gordon and Phillips (77,78), using an assay based on the viscosity of pectin solutions, have reported extracellular pectinolytic activity by two strains of anaerobic fungi including *Neocallimastix* sp. More recently these workers have described an extracellular pectin lyase produced by *Neocallimastix* sp. (78). The enzyme was shown to be endo-acting, with highest activity occurring at pH 8.0. The enzyme had catalytic requirement for $CaCl_2$ and was induced by pectin. Polygalacturonase, the other major enzyme of pectin degradation, has not yet been described in detail in anaerobic fungi.

E. Aryl Esterases

Plant cell-walls contain lignin that is bonded to hemicellulose by ester and ether linkages. Feruloyl esterase and *p*-coumaroyl esterase activities have been reported in culture filtrates from strains of *Piromyces*, *Neocallimastix*, and *Orpinomyces* (79). These enzymes are believed to cleave feruloyl and *p*-coumaroyl ester bridges linking lignin to the structural carbohydrates in the plant cell-wall, and they could be important in penetration of the plant structures by the fungus, providing maximal access to the polysaccharides. An extracellular *p*-coumaroyl esterase has been purified from the culture fluid of the *Neocallimastix* strain MC-2, which releases *p*-coumaroyl groups from *O*-[5-*O*-((*E*)-*p*-coumaroyl)-α-L-arabinofuranosyl]-(1-3)-*O*-β-D-xylopyranosyl-(1-4)-D-xylopyranose (PAXX). The enzyme has a molecular weight of 11 kDa under nondenaturing conditions (80). Under denaturing conditions the molecular weight is 5.8 kDa, suggesting that the enzyme may exist as a dimer under natural conditions. The isoelectric point of the enzyme is 4.7, and the pH optimum is 7.2. The purified enzyme releases *p*-coumaroyl groups from finely ground plant cells, and the activity is enhanced by the addition of xylanase and other plant cell-wall–degrading enzymes. Two feruloyl esterases have also been purified from the culture fluid of strain MC-2 (81). These enzymes cleave ferulic acid from *O*-{5-*O*-[(*E*)-feruloyl]-α-L-arabinofuranosyl}-(1-3)-*O*-β-D-xylopyranosyl-(1-4)-D-xylopyranose (FAXX), and have molecular weights of 69 and 24 kDA under both denaturing and nondenaturing conditions. One enzyme (FAE-1) is specific for FAXX, but the other (FAE-2) hydrolzes both FAXX and PAXX.

Mechanisms for the regulation of feruloyl and *p*-coumaroyl esterase production by anaerobic fungi are not yet known, but higher activities of these enzymes are obtained after growth on ground plant cell-walls than with purified substrates such as Avicel or xylan (79), indicating that they are inducible.

It appears that the inducer effect may be exerted through the complex ester linkages present in the plant material.

IV. REGULATORY MECHANISMS FOR SIMPLE SUGAR UPTAKE BY ANAEROBIC FUNGI

Anaerobic fungi are able to utilize a variety of simple sugars for growth, although some sugars such as L-arabinose, D-galactose, D-mannose, D-glucuronic acid, and D-galacturonic acid do not support growth (16,82). Enzymes for the utilization of disaccharides such as cellobiose, maltose, and xylobiose have already been described (Section III), and it appears that the highest activities of any one of these occur following growth on the structurally related oligomer or polysaccharide.

Regulatory mechanisms for simple sugar uptake similar to those found in rumen bacteria (83,84) occur in anaerobic fungi. Mountfort and Asher (82) showed that with paired-substrate test systems *N. frontalis* preferentially utilizes glucose as compared with fructose and xylose. The disaccharides cellobiose and sucrose are also preferentially utilized as compared with fructose and glucose, respectively, and cellobiose is the substrate preferred to xylose. Growth of the fungus on the substrate utilized last in the paired-substrate tests, followed by the addition of the preferred substrate and subsequent inhibition of utilization of the nonpreferred substrate, is indicative of catabolite regulation. Under these conditions utilization of both fructose and xylose is suppressed by glucose. The nature of catabolite regulation of simple sugar uptake by *N. frontalis* has not yet been elucidated, and it is not known whether inhibition, repression, or some other mechanism is involved.

V. CONCLUSIONS

Anaerobic fungi display a broad substrate specificity and are capable of utilizing the common polysaccharides present in plant cell-walls together with a variety of simple sugars. The utilization of carbohydrates by these organisms can in some situations be under regulatory constraint. Glucose and pentoses at elevated levels are able to exert a regulatory effect on the utilization of plant polysaccharides by regulating the synthesis of the enzymes involved in their degradation (Section III). Glucose can also regulate the utilization of simple sugars such as xylose and fructose (Section IV). These effects are summarized in Table 2. Yet their significance in the rumen ecosystem is unclear. In the rumen the situation almost continually exists where simple sugar substrates are exhausted because of intense competition between the microflora for their utilization (85), although under certain conditions the levels may rise, as at certain stages in the feeding cycle or with high-starch diets. However, it is questionable whether these sugars would ever reach the levels of 20 mM required to repress polysaccharidases in anaerobic fungi. It is not clear whether catabolite regulation of simple sugar utilization in anaerobic fungi, such as that described for *N. frontalis*, would in fact operate in the rumen either. In other microorganisms it is well known that similar regulatory mechanisms function only when the regulating sugar is at elevated concentrations and are lost when the concentrations are reduced (86,87).

The mechanisms for regulation of carbohydrate utilization in anaerobic fungi have probably evolved to give these organisms a competitive advantage under certain conditions in anaerobic environments. It is probable that the repressor mechanisms for polysaccharidases developed in situations where there was sufficient soluble sugar to satisfy much of the growth need of the fungus, allowing the less accessible polysaccharides to be conserved until such time as the soluble substrate had been exhausted. The mechanism would also function to prevent polysaccharide from becoming freely available to potential competitors that contained no polysaccharidase enzymes. It is of interest, however, that in anaerobic fungi, some polysaccharidases, namely cellulases, are constitutive (Section III.A.1). This property may provide them with a competitive advantage in situations where inducible cellulases might otherwise be repressed, as during the period of initial fungal colonization in the rumen, where soluble carbohydrates, particularly glucose, may be present at relatively high concentrations. The constitutive cellulases also have a broad substrate specificity in being able to attack crystalline and amorphous forms of cellulose in addition to xylans. Thus, these enzymes would provide the fungi with a competitive advantage in facilitating the rapid colonization and penetration of plant tissues, overcoming the delays that would otherwise occur because of repression of inducible enzymes.

Probably the major regulatory influences on carbohydrate fermentation by anaerobic fungi in the rumen are due to the levels of fermentation end products, and the low partial pressure of H_2 in this ecosystem (85) would most likely be the major influence. This would favor the production of more oxidized products such as acetate, and fungal growth would be optimized because of increased energy yields expected from the "acetate" pathway. Under such conditions electron-sink products such as lactate and ethanol would not be expected to exert any influence because of their low levels. An exception to this situation could arise if the animal were fed a diet high in soluble sugars such as those in beets or whey, resulting in acidosis. Under such conditions lactate would accumulate and might be expected to exert a negative-feedback effect on fermentation (Section II). Carbon dioxide in the rumen would also be expected to exert an effect on some enzymes involved in fungal fermentations, such as pyruvate:ferredoxin oxidoreductase (37), and because of the repressive effect of CO_2 on this enzyme, acetate formation would be expected to occur via a non-ferredoxin-linked pathway, and reducing power for H_2 via pyridine nucleotide:ferredoxin oxidoreductase. Perturbation of fungal fermentations because of the presence of other microbial fermentation end products in the rumen has not yet been studied, and the influence of alternative hydrogen acceptors to methane such as nitrate, sulfate, and fumarate (88) also remains to be investigated.

REFERENCES

1. Liebetanz E. Die parasitischen Protozoen des Widerkäuermagens. Arch Protistenks 1910; 19:19.
2. Braune R. Untersuchungen über die im Widerkäuermagen vorkommenden Protozoen. Arch Protistenks 1913; 32:111–170.
3. Barr DJS, Kudo H, Jakober KD, Cheng KJ. Morphology and development of rumen fungi: *Neocallimastix* sp., *Piromyces communis*, and *Orpinomyces bovis* gen. nov., sp. nov. Can J Bot 1989; 67:2815–2824.
4. Bauchop T, Mountfort DO. Cellulose fermentation by a rumen anaerobic fungus in both the absence and the presence of rumen methanogens. Appl Environ Microbiol 1981; 42:1103–1110.
5. Borneman WS, Akin DE, Ljungdahl LG. Fermentation products and plant cell wall degrading enzymes produced by monocentric and polycentric anaerobic ruminal fungi. Appl Environ Microbiol 1989; 55:1066–1073.
6. Breton A, Bernalier A, Bonnemoy F, et al. Morphological and metabolic characterization of a new species of strictly anaerobic rumen fungus: *Neocallimastix joyonii*. FEMS Microbiol Lett 1989; 58:309–314.
7. Breton A, Bernalier A, Dusser M. et al. *Anaeromyces mucronatus* nov. gen., nov. sp. A new strictly anaerobic rumen fungus with polycentric thallus. FEMS Microbiol Lett 1990; 70:177–182.
8. Ho YW, Bauchop T, Abdullah N, Jaladin S. *Ruminomyces elegans* gen et sp nov, a polycentric anaerobic rumen fungus from cattle. Mycotaxon 1990; 38: 397–405.
9. Kostyukovsky VA, Okenuv ON, Tarakanov BV. Description of two anaerobic fungal strains from the bovine rumen and influence of diet on fungal population *in vivo*. J Gen Microbiol 1991; 137:1759–1764.
10. Joblin KN. Isolation, enumeration and maintenance of rumen anaerobic fungi in roll-tubes. Appl Environ Microbiol 1981; 42:1119–1122.
11. Lowe SE, Theodorou MK, Trinci APJ, Hespell RB. Growth of anaerobic rumen fungi on defined and semi-defined media lacking rumen fluid. J Gen Microbiol 1985; 131:2225–2229.
12. Orpin CG. Studies on the rumen flagellate *Neocallimastix frontalis*. J Gen Microbiol 1975; 91:249–262.
13. Orpin CG. Studies on the rumen flagellate *Sphaeromonas communis*. J. Gen Microbiol 1976; 94:270–280.
14. Orpin CG. The rumen flagellate *Piromonas communis*. Its life-history and invasion of plant materials in the rumen. J Gen Microbiol 1977; 99:107–117.

15. Orpin CG, Munn EA. *Neocallimastix patriciarum* sp. nov., a new member of the Neocallimasticaceae inhabiting the rumen of sheep. Trans Br Mycol Soc 1986; 86:178-181.
16. Phillips MW, Gordon, GLR. Sugar and polysaccharide fermentation by rumen anaerobic fungi from Australia, Britain and New Zealand. BioSystems 1988; 21:377-383.
17. Gold JJ, Heath IB, Bauchop T. Ultrastructural description of a new chytrid genus of caecum anaerobe, *Caecomyces equi* gen. nov., assigned to the Neocallimasticaceae. BioSystems 1988; 21:403-415.
18. Li J, Heath IB, Bauchop T. *Piromyces mae* and *Piromyces dumbonica*, two new species of uniflagellate anaerobic chytridiomycete fungi from the hindgut of the horse and elephant. Can J Bot 1990; 68:1021-1033.
19. Orpin CG. Isolation of cellulolytic phycomycete fungi from the caecum of the horse. J Gen Microbiol 1981; 123:287-296.
20. Bauchop T. Biology of gut anaerobic fungi. BioSystems 1989; 23:53-64.
21. Orpin CG. The occurrence of chitin in the cell walls of the rumen organisms *Neocallimastix frontalis*, *Piromonas communis*, and *Sphaeromonas communis*. J Gen Microbiol 1977; 99:215-218.
22. Gay L. Chitin content and chitin synthase activity as indicators of the growth of three different anaerobic rumen fungi. FEMS Microbiol Lett 1991; 80:99-102.
23. Bauchop T. The rumen anaerobic fungi: colonizers of plant fibre. Ann Rech Vet 1979; 10:246-248.
24. Bauchop T. Rumen anaerobic fungi of cattle and sheep. Appl Environ Microbiol 1979; 38:148-158.
25. Orpin CG. Invasion of plant tissue in the rumen by the flagellate *Neocallimastix frontalis*. J Gen Microbiol 1977; 98:423-430.
26. Bauchop T. The anaerobic fungi in rumen fiber digestion. Agric Environ 1981; 6:339-348.
27. Akin DE, Gordon GLR, Hogan JP. Rumen bacterial and fungal degradation of *Digitaria pentzii* grown with or without sulfur. Appl Environ Microbiol 1983; 46:738-748.
28. Orpin CG, Letcher AJ. Utilization of cellulose starch, xylan, and other hemicelluloses for growth by the rumen phycomycete *Neocallimastix frontalis*. Curr Microbiol 1979; 3:121-124.
29. Bernalier A, Fonty G, Bonnemoy F, Gouet P. Degradation and fermentation of cellulose by the rumen anaerobic fungi in axenic culture or in association with cellulolytic bacteria. Curr Microbiol 1992; 25:143-158.
30. O'Fallon JV, Wright RW, Calza RE. Glucose metabolic pathways in the anaerobic fungus *Neocallimastix frontalis*. Biochem J 1991; 274:595-599.
31. Teunisson MJ, Smits AAM, Op den Camp HJM, et al. Fermentation of cellulose and production of cellulolytic and xylanolytic enzymes by anaerobic fungi from ruminant and non-ruminant herbivores. Arch Microbiol 1991; 156:290-296.
32. Chen M, Wolin MJ. Influence of methane production by *Methanobacterium ruminantium* on the fermentation of glucose and lactate by *Selenomonas ruminantium*. Appl Environ Microbiol 1977; 34:756-759.

33. Latham MJ, Wolin MJ. Fermentation of cellulose by *Ruminococcus flavefaciens* in the presence and absence of *Methanobacterium ruminantium*. Appl Environ Microbiol 1977; 34:297–301.

34. Scheifinger CC, Linehan B, Wolin MJ. H_2 productin by *Selenomonas ruminantium* in the presence of methanogenic bacteria. Appl Microbiol 1975; 29:480–483.

35. Wolin MJ. Metabolic interactions among intestinal microorganisms. Am J Clin Nutr 1974; 27:1320–1328.

36. Heath IB, Bauchop T, Skipp RA. Assignment of the rumen anaerobe *Neocallimastix frontalis* to the Spizellomycetales (Chytridiomycetales) on the basis of its polyflagellate zoospore ultrastructure. Can J Bot 1983; 61:295–307.

37. Yarlett N, Orpin CG, Munn EA, et al. Hydrogenosomes in the rumen fungus *Neocallimastix patriciarum*. Biochem J 1986; 236:729–739.

38. Yarlett N, Coleman GS, Williams AG, Lloyd D. Hydrogenosomes in known species of rumen entodiniomorph protozoa. FEMS Microbiol Lett 1984; 21:15–19.

39. Yarlett N, Hann AC, Lloyd D, Williams AG. Hydrogenosomes in the rumen protozoan *Dasytricha ruminantium* Schuberg. Biochem J 1981; 200:365–372.

40. Mountfort DO, Asher RA, Bauchop T. Fermentation of cellulose to methane and carbon dioxide by a rumen anaerobic fungus in a triculture with *Methanobrievibacter* sp. strain RA-1 and *Methanosarcina barkeri*. Appl Environ Microbiol 1982; 44:128–134.

41. Joblin KN, Naylor GE, Williams AG. Effect of *Methanobrevibacter smithii* on the xylanolytic activity of anaerobic ruminal fungi. Appl Environ Microbiol 1990; 56:2287–2295.

42. Richardson AJ, Stewart CS, Campbell GP, et al. Influence of coculture with rumen bacteria on the lignocellulolytic activity of phycomycetous fungi from the rumen. Proceedings of the XIV International Congress of Microbiology 1986; Abstract PG2–24.

43. Joblin KN, Naylor GE. Inhibition of the rumen anaerobic fungus *Neocallimastix frontalis* by fermentation end products. Lett Appl Microbiol 1993; 16:254–256.

44. Barichievich EM, Calza RE. Media carbon induction of extracellular cellulase activities in *Neocallimastix frontalis* isolate EB 188. Curr Microbiol 1990; 20:265–271.

45. Barichievich EM, Calza RE. Supernatant protein and cellulase activities of the anaerobic ruminal fungus *Neocallimastix frontalis* isolate EB 188. Appl Environ Microbiol 1990; 56:43–48.

46. Calza RE. Regulation of protein and cellulase excretion in the ruminal fungus *Neocallimastix frontalis* EB 188. Curr Microbiol 1990; 21:109–115.

47. Mountfort DO, Asher RA. Production and regulation of cellulase by two strains of the rumen anaerobic fungus *Neocallimastix frontalis*. Appl Environ Microbiol 1985; 49:1314–1322.

48. Williams AG, Orpin CG. Polysaccharide-degrading enzymes formed by three species of anaerobic rumen fungi grown on a range of carbohydrates. Can J Microbiol 1987; 33:418–426.

49. Wilson CA, Wood TM. Studies on the cellulase of the rumen anaerobic fungus *Neocallimastix frontalis*, with special reference to the capacity of the enzyme to degrade crystalline cellulose. Enz Microb Technol 1992; 14:258-264.

50. Wood TM, Wilson CA, McCrae SI, Joblin KN. A highly active extracellular cellulase from the anaerobic rumen fungus *Neocallimastix frontalis*. FEMS Microbiol Lett 1986; 34:37-40.

51. Li X, Calza RE. Fractionation of cellulases from the ruminal fungus *Neocallimastix frontalis* EB 188. Appl Environ Microbiol 1991; 57:3331-3336.

52. Wallace RJ, Joblin K. Proteolytic activity of a rumen anaerobic fungus. FEMS Microbiol Lett 1985; 29:19-25.

53. Wallace RJ, Munro CA. Influence of the rumen anaerobic fungus *Neocallimastix frontalis* on the proteolytic activity of a defined mixture of rumen bacteria grown on a solid substrate. Lett Appl Microbiol 1986; 3:23-26.

54. Li X, Calza RE. Cellulases from *Neocallimastix frontalis* EB 188 synthesized in the presence of glycosylation inhibitors: measurement of pH and temperature optima, protease and ion sensitivities. Appl Microbiol Biotechnol 1991; 35:741-747.

55. Xue GP, Orpin CG, Gobius KS, et al. Cloning and expression of multiple cellulase cDNAs from the anaerobic fungus *Neocallimastix frontalis* in *E. coli*. J Gen Microbiol 1992; 1338:1413-1420.

56. Xue GP, Gobius KS, Orpin CG. Isolation of a multifunctional cellulase (*cel E*) from the rumen fungus *Neocallimastix patriciarum*. Proceedings of the XVII International Grasslands Congress. Palmerston North, New Zealand, Vol 2, 1993, p 1221-1222.

57. Xue GP, Gobius KS, Orpin CG. A novel polysaccharide hydrolase cDNA (*cel D*) from *Neocallimastix patriciarum* encoding three multifunctional catalytic domains with high endoglucanase, cellobiohydrolase, and xylanase activities. J Gen Microbiol 1993; 138:2397-2403.

58. Alward JH, Xue GP, Simpson GO, Orpin CG. Cellobiohydrolase (CBH) from *Neocallimastix patriciarum*: a membrane associated complex? Proc XVII International Grasslands Congress. Palmerston North, New Zealand, Vol 2, 1993, p 1222-1224.

59. Williams AG, Orpin CG. Glycoside hydrolase enzymes present in the zoospore and vegetative growth stages of the rumen fungi *Neocallimastix patriciarum*, *Piromonas communis*, and an unidentified isolate, grown on a range of carbohydrates. Can J Microbiol 1987; 33:427-434.

60. Li X, Calza RE. Purification and characterization of an extracellular β-glucosidase from the rumen anaerobic fungus *Neocallimastix frontalis* EB 188. Enz Microb Technol 1991; 13:1-7.

61. Li X, Calza RE. Kinetic study of a cellobiase purified from *Neocallimastix frontalis* EB 188. Biochim Biophys Acta 1991; 1080:148-154.

62. Teunissen MT, Lahaye DHTP, Huis in't Veld JHJ, Vogels GD. Purification and characterization of an extracellular β-glucosidase from *Piromyces* sp. strain E2. Arch Microbiol 1992; 158:276-281.

63. Teunissen MJ, De Kort GVM, Op den Camp HJM, Huis in't Veld JHJ. Production of cellulolytic and xylanolytic enzymes during the growth of the anaero-

bic fungus *Piromyces* sp. on different substrates. J. Gen Microbiol 1992; 138: 1657–1664.

64. Theodorou MK, Longland AC, Dhanoa MS, et al. Growth of *Neocallimastix* sp. str. R1 on Italian rye-grass hay: removal of neutral sugars from plant cell walls. Appl Environ Microbiol 1989; 55:1363–1367.

65. Gilbert HJ, Hazlewood GP, Laurie JI, et al. Homologous catalytic domains in a rumen fungal xylanase: evidence for gene duplication and procaryotic origin. Mol Microbiol 1992; 6:2065–2072.

66. Lowe SE, Theodorou MK, Trinci APJ. Cellulases and xylanase of an anaerobic rumen fungus grown on wheat straw, wheat straw holocellulose, cellulose and xylan. Appl Environ Microbiol 1987; 53:1216–1223.

67. Mountfort DO, Asher RA. Production of xylanase by the ruminal anaerobic fungus *Neocallimastix frontalis*. Appl Environ Microbiol 1989; 55:1016–1022.

68. Teunissen MJ, Hermans JMH, Huis in't Veld JHJ, Vogels GD. Purification and characterization of a complex bound and free β-1,4-endoxylanase from the culture fluid of the anaerobic fungus *Piromyces* sp. strain E2. Arch Microbiol 1993; 159:265–271.

69. Garcia Campayo V, Wood TM. Purification and characterization of a β-D-xylosidase from the rumen anaerobic fungus *Neocallimastix frontalis*. Carbohydr Res 1993; 242:229–245.

70. Hébraud M. Fèvre M. Purification and characterization of extracellular β-xylosidase from the rumen anaerobic fungus *Neocallimastix frontalis*. FEMS Microbiol Lett 1990; 72:11–16.

71. Slyter LL. Influence of acidosis on rumen function. J Anim Sci 1976; 43:910–929.

72. Hobson PN, MacPherson M. Amylase from *Clostridium butyricum* and a streptococcus isolated from the rumen of the sheep. Biochem J 1952; 52:671–679.

73. Prins RA. Biochemical activities of gut micro-organisms. In: Clarke RTJ, Bauchop T, eds. Microbial Ecology of the Gut. London: Academic Press, 1977:73–183.

74. Walker GJ. The cell-bound α-amylase of *Streptococcus bovis* sp. Biochem J 1965; 94:289–298.

75. Walker GJ, Hope PM. Degradation of starch granules by some amylolytic bacteria from the rumen of sheep. Biochem J 1964; 90:398–408.

76. Mountfort DO, Asher RA. Production of α-amylase by the ruminal anaerobic fungus *Neocallimastix frontalis*. Appl Environ Microbiol 1988; 54:2293–2299.

77. Gordon GLR, Phillips MW. Pectin degradation by some ruminal anaerobic fungi. In: Zahari MW, Tajuddin ZA, Abdullah N, Wong HK, eds. Proceedings of the 3rd International Symposium on Nutrition of Herbivores. Serdang. Malaysian Society of Animal Production, 1991:27.

78. Gordon GLR, Phillips MW. Extracellular pectin lyase produced by *Neocallimastix* sp, LM1, a rumen anaerobic fungus. Lett Appl Microbiol 1992; 15:113–115.

79. Borneman WS, Hartley RD, Morrison WH, et al. Feruloyl and *p*-coumaroyl esterase from anaerobic fungi in relation to plant cell wall degradation. Appl Microbiol Biotechnol 1990; 33:345–351.

80. Borneman WS, Ljungdahl LG, Hartley RD, Akin DE. Isolation and characterization of p-coumaroyl esterase from the anaerobic fungus *Neocallimastix* Strain MC-2. Appl Environ Microbiol 1991; 57:2337–2344.

81. Borneman WS, Ljungdahl LG, Hartley RD, Akin DE. Purification and partial characterization of two feruloyl esterases from the anaerobic fungi *Neocallimastix* Strain MC-2. Appl Environ Microbiol 1992; 58:3762–3766.

82. Mountfort DO, Asher RA. Role of catabolite regulatory mechanisms in control of carbohydrate utilization by the rumen anaerobic fungus *Neocallimastix frontalis*. Appl Environ Microbiol 1983; 46:1331–1338.

83. Russell JB, Baldwin RL. Substrate preferences in rumen bacteria: evidence of catabolite regulatory mechanisms. Appl Environ Microbiol 1978; 36:319–329.

84. Russell JB, Baldwin RL. Comparison of substrate affinities among several rumen bacteria: a possible determinant of rumen bacterial competition. Appl Environ Microbiol 1979; 37:531–536.

85. Clarke RTJ. The gut and its microorganisms. In: Clarke RTJ, Bauchop T, eds. Microbiol Ecology of the Gut. London: Academic Press, 1977: 36–65.

86. Mateles RI, Chian SK, Silver R. Continuous culture on mixed substrates. In: Powell EO, Evans CGT, Strange RE, Tempest DW, eds. Microbial Physiology and Continuous Culture. London: Her Majesty's Stationery Office, 1967:232–239.

87. Silver RS, Mateles RI. Control of mixed substrate utilization in continuous cultures of *Escherichia coli*. J Bacteriol 1969; 97:535–543.

88. Wolin MJ. Hydrogen transfer to microbial communities. In: Bull AT, Slater JH, eds. microbial Interactions and Communities. Vol 1. London: Academic Press, 1982:323–356.

6

Ultrastructure
of Plant Cell-Walls Degraded by Anaerobic Fungi

DANNY E. AKIN

Richard B. Russell Agricultural Research Center
Agricultural Research Service
U.S. Department of Agriculture
Athens, Georgia

I. INTRODUCTION

Anaerobic fungi indigenous to the rumen are a part of the fiber-degrading population in this ecosystem (1,2,3). Because of the unusual nature of these organisms and the lack of appropriate guidelines for classification, there has been some uncertainty about the taxonomy of cultures, and possibly some names will be modified at a later date. In this chapter genera are designated according to Barr et al. (4), and their characteristics are as follows: *Neocallimastix*, monocentric growth, multiflagellate zoospores; *Piromyces*, monocentric growth, uniflagellate zoospores; *Orpinomyces*, polycentric growth, multiflagellate zoospores.

The ability of anaerobic fungi to degrade and utilize structural carbohydrates from plant cell-walls has been investigated in isolates of these microorganisms representing a wide geographical distribution including Australia (5), Great Britain (6–9), France (10), New Zealand (11,12,13), and the United States (14). Isolates from these various countries are similar in that they produce high amounts of extremely active cellulases and xylanases. For example, Wood et al. (9) reported that a unit of *N. frontalis* endoglucanase (from cell-free preparations of fungus–methanogenic bacterium coculture) was several times more active than that from *Trichoderma reesei* C-30, which is considered to be one of the most active cellulases reported to date. Further,

169

anaerobic fungi are extremely active in producing xylanases. Mountfort and Asher (12) reported that xylanase produced by *N. frontalis* was one of the most active of the endo-acting polysaccharides yet studied from the anaerobic fungi. Pectin is one plant cell-wall polysaccharide, however, that does not appear to be degraded or utilized (5,6,7,15). Cell-wall carbohydrases from several genera, as reported by Borneman et al. (14), are shown in Table 1.

In studies of the ultrastructure of plant cell-wall degration, several methods have been used to evaluate the activity of the anaerobic fungi. Removal of fragments of forage material directly from the rumen (i.e., the in vivo method) allows an evaluation of the activity of the mixed fungal population without manipulations within the laboratory; however, in this method the exact time of adherence to fiber, the state of the plant material, and the influence of other microorganisms in the cosystem cannot be precisely determined. In vitro incubation of plant fragments allows precise definition of plant substrates, and the use of antibiotics against specific microbial groups allows evaluation of plant wall degradation by procaryotes or eucaryotes. In our system (16,17), fiber degradation by eucaryotes is carried out mostly by anaerobic fungi and not rumen protozoa. On the basis of studies by Bauchop and Mountfort (18), we have successfully used streptomycin (130 U ml^{-1} of broth or rumen fluid) plus penicillin (2000 U ml^{-1}) to inhibit fiber-degrading bacteria, and chloramphenicol (30 μg ml^{-1}) to inhibit other groups such as methanogenic bacteria. By including cycloheximide (0.5 mg ml^{-1}) in the medium or rumen fluid, the fungi can be inhibited and the potential activity of bacteria can be evaluated. This in vitro method permits an evaluation of the potential activity of the mixed fungal population but not of the actual activity when the fungi must compete with other microorganisms in the rumen

Table 1 Enzyme activities of Anaerobic Fungi Against Plant Cell Wall-Type Polysaccharides

	Specific activity (U mg^{-1} supernatant protein)[a]				
Fungus	Exoglucanase	Endoglucanase	β-Glucosidase	Xylanase	β-Xylosidase
Piromyces MC-1	0.16	2.03	1.32	4.51	1.32
Neocallimastix MC-2	0.20	2.83	1.68	14.45	1.48
Orpinomyces PC-2	0.12	3.01	2.08	20.16	0.81
Orpinomyces PC-3	0.11	3.04	1.82	15.48	0.70
Anaeromyces PC-1	0.12	2.58	1.93	12.97	0.43

[a]Maximum specific activity within 9 days growth on Bermuda grass assayed against the following substrates: Avicel, carboxymethylcellulose, *p*-nitrophenyl β-D-glucopyranoside, xylan (larchwood), *p*-nitrophenyl β-D-xylopyranoside.
Source: Data from Borneman et al. (14).

ecosystem (19). A third method used to evaluate the cell wall–degrading activity of rumen fungi is axenic culture. Fungi have been isolated and purified to axenic cultures by growth of single zoospores in the presence of antibacterial antibiotics (18). This method allows for a direct comparison of individual organisms based on growth rates as well as their rate and extent of plant wall degradation due to variations in the amount or types of enzymes. Our studies have shown that all three methods indicate similar ultrastructural patterns of fungal attack on plant tissues as substrates, although rates and extents could vary among methods.

II. PLANT STRUCTURE AND RELATIONSHIP TO BIODEGRADATION

The structural carbohydrates cellulose and hemicellulose, which are the prevalent polysaccharides in most forage cell-walls, are virtually totally available for biodegradation if not limited by other factors (20). However, the covalent linkage of phenolics (e.g., lignins) to the cell-wall carbohydrates and subsequent "protection" of polysaccharides from microbial enzymes inhibit the utilization of an otherwise available substrate for growth of rumen microorganisms (21,22). The leaf blades (Fig. 1) and stems (Fig. 2) of plants are composed of tissues having various arrangements and cell-wall thickness, with the phenolic compounds (e.g., lignin) concentrated in particular cell-wall types. Lignin provides rigidity and mechanical support for plants and, consequently, results in reduced fiber biodegradation. Microscopic studies using histochemical stains for lignin have provided information on the anatomy and structure of plants used as substrates for microorganisms (Table 2).

The tissues giving the most intense and consistent reactions for lignin and composing most of the plant residue after digestion are the xylem, mestome sheath, and sclerenchyma in leaf blades and the epidermis, sclerenchyma ring, vascular tissue, and mature parenchyma in stems. In addition to intense histochemical reacitons for lignin (Table 2), other studies using ultraviolet absorption microspectrophotometry (23,24) of individual cell walls have confirmed that these tissues have the highest concentrations of phenolics.

Tissues giving different reactions to acid phloroglucinol and chlorine-sulfite stains show different reponses to biodgradation (Table 2) and to chemical treatment. The chlorine sulfite tissues (e.g., sclerenchyma in grass leaf blades and parenchyma in stems) are rendered more readily available for microbial degradation after alkali treatment than are the acid phloroglucinol tissues (e.g., sclerenchyma ring in stems) (25,26). Furthermore, the former tissues are "delignified" completely after treatment with potassium permanganate, based on fluorescence of cell-wall phenolics, while this treatment causes the latter tissues to lose only about 30% of the fluorescence (27). Therefore, while both chlorine sulfite- and acid phloroglucinol–positive tissues

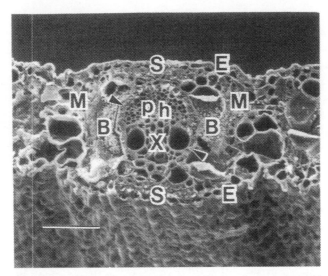

Figure 1 Scanning electron micrograph of a leaf cross section of Bermuda grass showing the arrangement of tissue. S = sclerenchyma; X = xylem; B = parenchyma bundle sheath; ph = phloem; M = mesophyll; E = epidermis. Arrows point to the mestome sheath. Scale bar = 50 μm.

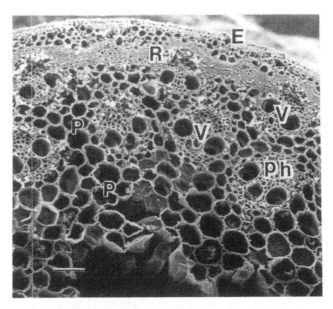

Figure 2 Scanning electron micrograph of an internode cross section of Bermuda grass showing the arrangement of tissues. E = epidermis; R = sclerenchyma ring; V = vascular bundle; ph = phloem; P = parenchyma. Scale bar = 50 μm.

Table 2 Characteristics of Tissue Types in Forage Grasses

Tissue	% Cross-sectional area	Lignin reaction[a]	Digestibility[b]
Grass leaf blade			
Xylem and mestome sheath	2–10	AP + +	None
Sclerenchyma	3–11	CS + +	Partial
Parenchyma bundle sheath	9–28	CS + ; none	Complete; partial
Epidermis	17–49	CS + ; none	Complete; partial
Mesophyll	33–65	None	Complete; rapid
Grass stem			
Epidermis	1–6	AP + +	None
Sclerenchyma ring	13–23	AP + +	None
Vascular tissue	7–14	AP + +	None
Parenchyma	48–62		
Immature		None	Total
Mature		AP + ; CS +	Partial; none

[a]Histological reaction for lignin with acid phloroglucinol (AP) or chlorine sulfite (CS); + +, + = intensity of reaction.
[b]Relative digestibility when sections are exposed to fiber-degrading rumen bacteria.
Source: Data from Akin (42) (compiled from literature).

have high concentrations of phenolics, inherent chemical or structural factors render the acid phloroglucinol–positive tissues the more resistant to biodegradation.

III. FUNGAL COLONIZATION OF PLANT TISSUES

Anaerobic fungi readily colonize plant cell-walls, including the more recalcitrant and lignified tissues of plants. In Fig. 3 *Neocallimastix* MC-2, which is a monocentric organism, is shown colonizing the cut edge of a leaf blade. Initially, zoospores swim to acceptable sites on the plant, encyst, and produce rhizoids that invade and penetrate the plant cell-walls (Fig. 3a). The fungus further develops and produces sporangia, which are attached to the plant substrate by rhizoids, and degrades the plant tissues (Fig. 3b). Sporangia are also produced on the leaf surface, where the rhizoids penetrate the cuticle barrier of epidermal cells (Fig. 3c). Colonization of plant substrates by fungi having polycentric growth differs in some ways from that by monocentric organisms. With polycentric fungi, often a rhizomycelium, which differs from a rhizoid by having nuclei and the capability of producing fruiting bodies (4), overgrows the exposed tissues of the plant (Fig. 4a). Production

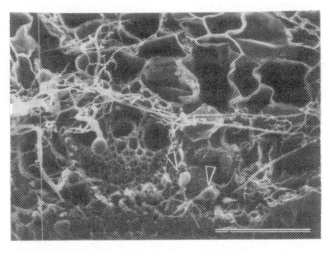

(a)

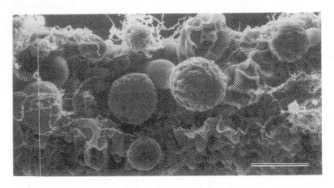

(b)

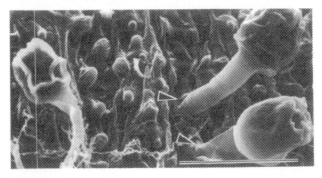

(c)

of filaments in these organisms can be extensive, and often sporangia are not produced. Prolific growth through the plant tissues can occur, resulting in degradation of cell walls and disruption of the cuticle barrier (Fig. 4b). Invasion of plant walls by filments results in degradation of unlignified tissues, such as mesophyll and epidermis, and partial degradation of the more recalcitrant tissues (Fig. 4c). Results from in vivo, in vitro, and axenic culture studies have all shown that anaerobic fungi frequently and extensively colonize the lignified tissues (2,28); fungi often preferentially associate with leaf blade sclerenchyma (1).

IV. DEGRADATION OF PLANT TISSUES

The colonization of exposed areas of plants by anaerobic fungi is accompanied by invasion of tissues by rhizoids or rhizomycelia, resulting in degradation of plant walls. The mixed fungal population of the rumen readily degrades forages of various types, digesting a substantial amount of fiber in intact sections of the morphological parts of plants (Table 3). While most plant parts show extensive degradation of plant walls (i.e., 30% or more), some parts, such as leaves of legumes, are not degraded as well. In the rumen, leafy diets are reported to result in a reduced fungal population (2), possibly because of rapid passage of such forage through the digestive tract and lack of sufficient time for colonization. However, digestibility coefficients in Table 3 from in vitro studies suggest that leaves per se of at least some legumes are poorly degraded by a mixed fungal population. In the case of *Lespedeza* the lack of digestion is probably due to the high level of tannins present in the leaves of this plant (29).

Scanning and transmission electron microscopy have been useful in delineating the ease and extent of degradation of specific tissues (1,2,28). Where fungal colonization is heavy on lignified tissues of an exposed area, digestion of nearby nonlignified tissues (e.g., mesophyll) is rapid and complete; however, such tissues remain undigested in regions where there is no colonization (Fig. 5). Rumen fungi predominantly colonize the sclerenchyma and vascular tissues of leaf blades, with filaments invading the plant cell-walls

---

Figure 3 Scanning electron micrographs of leaf cross sections of Bermuda grass colonized by *Neocallimastix* MC-2. Scale bar = 50 μm. (a) Young sporangia (arrows) with rhizoids are present particularly on the sclerenchyma after 24 h incubation, but plant cell-wall degradation is negligible. (b) Sporangia are present after 72 h incubation, and cell-wall degradation is extensive on cut end of the blade. (c) Colonization of the leaf surface and penetration of the cuticle by stalked sporangia (arrows) at 168 h incubation. (From Ref. 41.)

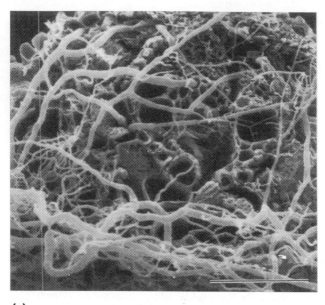

(a)

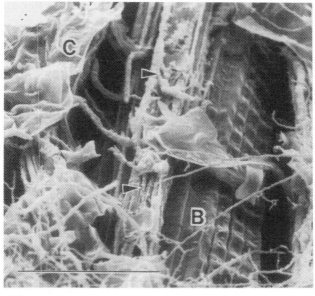

(b)

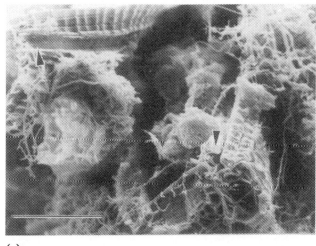

(c)

Figure 4 Scanning electron micrographs of leaf cross sections of Bermuda grass colonized by polycentric fungi. Scale bar = 50 μm. **(a)** Orpinomyces PC-3 extensively overgrowing the cut edge and penetrating the tissues without extensive degradation of plant cell-walls at 24 h. **(b)** *Anaeromyces* PC-1 rupturing the cuticle (C) over a large vascular bundle (B) and extensively degrading the underlying sclerenchyma cell walls (arrows) at 168 h incubation. **(c)** Degradation and rupture of mestome sheath (arrows) by PC-1 at 168 h incubation.

(Fig. 5). These filaments anchor the fungus to the recalcitrant tissues and erode the plant cell-wall, apparently by enzymes released from the fungal cell-wall (i.e., extracellular polysaccharidases) (Fig. 6a). After further growth, greater erosion of plant walls occurs, resulting in eventual degradation of all sclerenchyma walls (Fig 6b). In contrast, rumen bacteria adhere to the sclerenchyma wall and degrade the peripheral regions, which results in only moderate to slight degradation of this tissue (Fig. 6c). Area calculations from micrographs of tissues remaining in leaf residue after digestion show that the fungal population degrades significantly more sclerenchyma than does the bacterial population (Table 4).

The mestome sheath and xylem of leaf blades are generally more resistant to biodegradation than the sclerenchyma (Table 2). While often the sclerenchyma is eventually totally degraded, the mestome sheath and xylem are invaded by fungal filaments and degraded in few sites, resulting in an only partially degraded plant wall (Fig. 4c). Transmission electron microscopy shows the penetration and partial degradation by individual filaments in in vitro studies and axenic cultures (Fig. 6d,e). In contrast to the fungi, rumen bacteria have not been shown to adhere to or degrade the mestome sheath and xylem cell-walls to any significant extent (Fig. 6f).

Stems are generally less digestible than leaves, and increased maturity reduces the digestibility of stems faster than that of leaves (30). This low stem digestibility is associated with a higher level of phenolics and lignin within stem compared with leaf cell-walls (31). The epidermis, sclerenchyma ring, and vascular tissue in stems are particularly high in phenolics, and these tis-

Table 3   Dry Weight Loss of Intact Forages by Rumen Fungi[a]

Plant	Plant part	% Dry Weight loss[b]
Warm-season grasses		
Cynodon dactylon (Bermuda grass)	Blade	37.6 ± 4.2
	Sheath	38.5 ± 1.2
	Stem	31.4 ± 1.7
Sorghum bicolor (sorghum)	Blade	36.5 ± 0.6
	Sheath	36.3 ± 1.3
	Stem	25.7 ± 0.4
Cool-season grasses		
Phalaris arundinacea (reed canary-grass)	Blade	28.6 ± 2.1
	Sheath	40.0 ± 7.4
	Stem	45.2 ± 0
Festuca arundinacea (tall fescue)	Blade	44.5 ± 2.5
	Sheath	48.2 ± 2.1
Legumes		
Medicago sativa (lucerne)	Leaves	14.7 ± 1.8
	Stems	25.6 ± 1.8
Lespedeza cuneata (sericea lespedeza)	Leaves	0.9 ± 1.1
	Stems	46.0 ± 1.0

[a]Incubated 72 hours with rumen fluid plus streptomycin and penicillin.
[b]Percentages are averages plus standard deviations of triplicate tubes. Values are corrected for acid-pepsin–soluble materials and for loss in uninoculated controls.

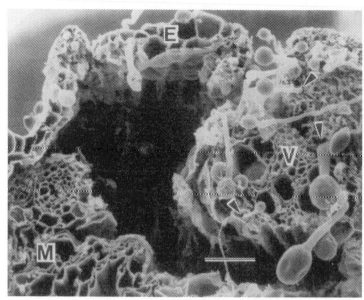

Figure 5 Scanning electron mirograph of leaf blade of reed canary-grass incubated for 72 h with rumen fluid containing streptomycin and penicillin. Fungi colonized the lignified vascular tissue (V), with rhizoids invading the plant structures (arrows). Digestible tissues, such as the mesophyll (M) and epidermis (E), are not degraded in regions not colonized. Scale bar = 50 μm.

Table 4 Comparative Degradation of Sclerenchyma Cell-Walls Incubated with Rumen Microorganisms for 48 Hours

	Area (μm^2) sclerenchyma tissue in residue	
Inoculum	Midvein	Second-order vein
Whole rumen fluid[a]	1543[c]	120[c]
Mixed rumen fungi[b]	356[d]	16[d]

[a]Activity mostly due to rumen bacteria.
[b]Rumen fluid plus streptomycin and penicillin to inhibit fiber-degrading bacteria and select rumen fungi.
[c,d]Values with different superscripts within a column differ, $p \leqslant 0.05$.
Source: Data from Akin and Rigsby (19).

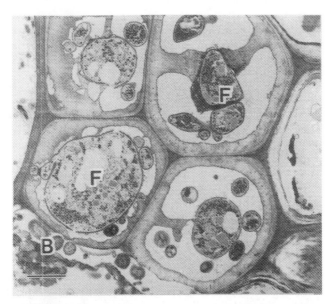

Figure 6a   Transmission electron microscopy of rumen fungal degradation of lignified tissues of leaf blades. Scale bar = 2 μm. **(a)** Sclerenchyma from leaf fragments removed directly from the bovine rumen showing the extensive degradation of plant walls near the fungal structures (F) and lack of degradation in regions near the bacteria (B). (From Ref. 19.)

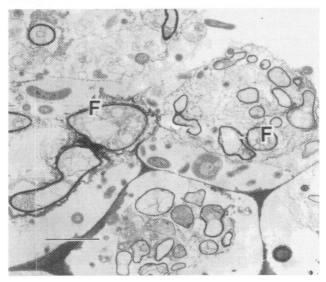

Figure 6b   Sclerenchyma digested in vitro showing complete degradation of plant walls by fungal structures (F). (From Ref. 42.)

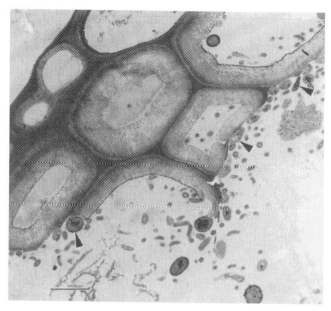

Figure 6c Sclerenchyma colonized by rumen bacteria (arrows) showing attack only at the periphery of the tissue. (From Ref. 42.)

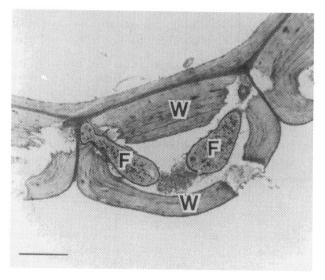

Figure 6d Mestome sheath incubated for 48 hours with rumen fluid containing streptomycin and penicillin showing partial degradation of the plant wall (W) by fungi (F).

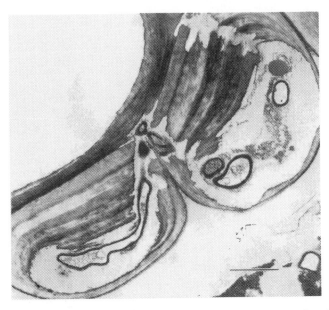

Figure 6e   Mestome sheath degraded by *Orpinomyces* PC-3 after incubation for
7 days.

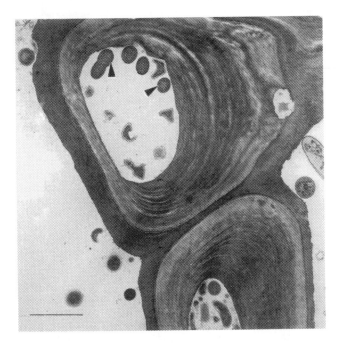

Figure 6f   Mestome sheath showing resistance to nearby rumen bacteria (arrows).

sues are the most resistant to biodegradation (Table 2). Rumen fungi colonize these tissues. In vivo studies show extensive sporangial development on sclerenchyma ring walls of stems after the plants have been masticated by the animal (Fig. 7a). In vitro studies indicate the potential for fungal colonization and attack on tissues in stems, showing rupture of vascular tissue (Fig. 7b). Transmission electron microscopy shows the growth across sclerenchyma ring walls (secondary, primary, and middle lamella layers) in stems by fungal filaments, with preferential attack in the pit fields in at least some plants (Fig. 8a). In contrast, rumen bacteria are able to slightly erode and cause pitting of these cell-wall types (Fig. 8b).

The high activity of fungal cellulases and hemicellulases is important in the degradation of recalcitrant plant walls. Additionally, Borneman et al. (32) recently reported for the first time in rumen fungi the presence of esterases that release *p*-coumaric and ferulic acids from ester-linked carbohydrate-phenolic complexes in plant walls. Such enzymes, in conjunction with the

Figure 7a Scanning electron micrograph of Bermuda grass stem colonized by rumen fungi. Scale bar = 50 μm. **(a)** Plant fragment from the rumen showing colonization of lignified tissues of an exposed area by sporangia on stalks.

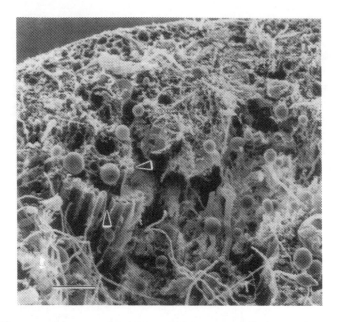

Figure 7b Incubation for 72 h with rumen fluid containing streptomycin and penicillin showing extensive colonization by fungi, invasion of lignified tissues by rhizoids, and partial degradation of the vascular cell walls (arrows).

active polysaccharidases, undoubtedly play an important role in fungal degradation of lignocellulose substrates.

Despite their ability to degrade lignin-containing plant walls, rumen fungi have not been shown to utilize phenolics or lignin. In studies using radiolabeled ^{14}C lignin or carbohydrate fractions, mixed rumen fungi (33) and axenic fungal cultures (5) solubilized but did not metabolize the phenolic moieties of plants. The addition of phenolic monomers at 10 = mM concentrations to broth medium reduced both fungal growth on plants and fiber degradation (19).

V. PHYSICAL WEAKENING OF PLANT RESIDUE

Flow of feed particles from the reticulorumen has been reported to be the major process controlling both intake and nutritive value of forages (34). Before feed particles exit the rumen, they must be reduced in size (35). Ingestive mastication and rumination by the animal are mostly responsible for reducing the size of particles, with activity by rumen microorganisms considered to contribute little (36). However, the contribution of prior microbial digestion to physical breakdown of particles, such as that which occurs in rumination, has not been well studied, and even less is known about the contribution of particular microbial groups such as the anaerobic fungi to this activity. The ability of the fungi to degrade the most recalcitrant plant walls and to penetrate and disrupt the cuticle suggests that their activity might allow for easier rumination of digesta. In studies (37) to compare quantitatively the abilities of rumen fungi and bacteria to weaken fiber, results showed that legume and grass stems incubated with fungi were significantly weaker than those incubated with bacteria in tests simulating the activity of chewing (i.e., shear strength) (Table 5). Axenic cultures of *Neocallimastix* MC-2 and *Orpinomyces* PC-2 weakened Bermuda grass internodes to a substantially greater degree than did axenic cultures of the main fiber-degrading rumen bacteria (i.e., *Fibrobacter succinogenes* and *Ruminococcus flavefaciens*) (38). This

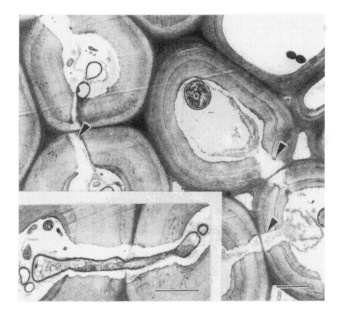

Figure 8a   Transmission electron micrograph of lignified stem tissue degraded by rumen microorganisms at 72 hours. Scale bar = 2 μm. **(a)** Incubation with rumen fluid containing penicillin and streptomycin showing penetration of cell walls (arrows). Inset shows a filament crossing two cells.

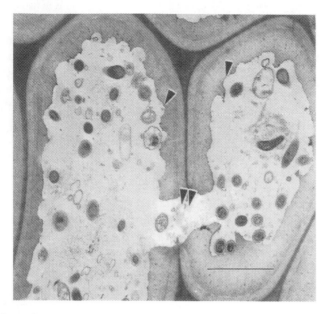

Figure 8b Incubation with rumen fluid plus cycloheximide showing slight pitting of the plant wall by rumen bacteria (arrows). (From Ref. 37.)

Table 5 Reduction in Textural Strength of Stem Internodes of Forages Incubated with Rumen Microorganisms for 72 Hours

	% Reduction in Stress (N/cm)[a]	
Inoculum	Lucerne	Bermuda-grass
Whole rumen fluid	9[d]	20[d,e]
Mixed rumen bacteria[b]	0[d]	30[e]
Mixed rumen fungi[c]	35[e]	57[f]
Control	3[d]	6[d]

[a]N = Newton (i.e., force that gives to a mass of 1 kg an acceleration of 1 m/s^2), divided by stem diameter in centimeters.
[b]Rumen fluid plus cycloheximide (0.5 mg ml^{-1} fluid) to inhibit eucaryotes and to select for rumen bcteria.
[c]Rumen fluid plus streptomycin (130 U ml^{-1} fluid) and penicillin (2000 U ml^{-1} fluid) to inhibit fiber-digesting bacteria and to select for fungi.
[d,e,f]Values with different superscripts within a column differ, $p \leqslant 0.05$.
Source: Data adapted from Akin et al. (37).

ability to reduce the textural strength of plants, while possibly a factor related to intake and passage of fiber, has not been evaluated in animal studies to date. However, in a comparison of ryegrass plants, lower values in leaf tensile strength have been related to a higher intake of forage (39). In animal studies of forage intake and fungal activity, the presence of a high fungal population has been positively correlated with residues weaker in tensile strength (1) and with chewing characterisitics related to better feed utilization (40).

VI. CONCLUSIONS

Anaerobic fungi have extremely active cellulases and hemicellulases that result in extensive degradation of plant cell walls. These enzymes, with phenolic acid esterases and perhaps with others not yet discovered, result in fungal attack and degradation of the structural carbohydrates in phenolic-containing plant cell-walls. The ability of anaerobic fungi to invade and penetrate the most recalcitrant plant walls and structural barriers to effect degradation and thereby to substantially weaken forage residues provides anaerobic fungi with unique attributes among microorganisms in the rumen ecosystem.

REFERENCES

1. Akin DE, Gordon GLR, Hogan JP. Rumen bacterial and fungal degradation of *Digitaria pentzii* grown with or without sulfur. Appl Environ Microbiol 1983; 46: 738–748.
2. Bauchop T. Rumen anaerobic fungi of cattle and sheep. Appl Environ Microbiol 1979; 38:148–158.
3. Orpin CG. Studies on the rumen flagellate *Neocallimastix frontalis*. J Gen Microbiol 1975; 91:249–262.
4. Barr DJS, Kudo H, Jakober KD, Cheng K-J. Morphology and development of rumen fungi: *Neocallimastix* sp., *Piromyces communis*, and *Orpinomyces bovis* gen. nov., sp. nov. Can J Bot 1989; 67:2815–2824.
5. Gordon GLR, Phillips MW. Comparative fermentation properties of anaerobic fungi from the rumen. In: Nolan JV, Leng RA, Demeyer DI, eds. The Rôles of Protozoa and Fungi in Ruminant Digestion. Armidale, Australia: Penambul Books, 1989:127–138.
6. Lowe SE, Theodorou MK, Trinci APJ. Cellulases and xylanase of an anaerobic rumen fungus grown on wheat straw, wheat straw holocellulose, cellulose, and xylan. Appl Environ Microbiol 1987; 53:1216–1223.
7. Williams AG, Orpin CG. Polysaccharide-degrading enzymes formed by three species of anaerobic rumen fungi grown on a range of carbohydrate substrates. Can J Microbiol 1987; 33:418–426.
8. Williams AG, Orpin CG. Glycoside hydrolase enzymes present in the zoospore and vegetative growth stages of the rumen fungi *Neocallimastix patriciarum*, *Piromonas communis*, and an unidentified isolate, grown on a range of carbohydrates. Can J Microbiol 1987; 33:427–434.

9. Wood TM, Wilson CA, McCrae SI, Joblin KN. A highly active extracellular cellulase from the anaerobic rumen fungus *Neocallimastix frontalis*. FEMS Microbiol Lett 1986; 34:37–40.

10. Hébraud M, Fèvre M. Characterization of glycoside and polysaccharide hydrolases secreted by the rumen anaerobic fungi *Neocallimastix frontalis*, *Sphaeromonas communis* and *Piromonas communis*. J Gen Microbiol 1988; 134:1123–1129.

11. Mountfort DO, Asher RA. Production and regulation of cellulase by two strains of the rumen anaerobic fungus *Neocallimastix frontalis*. Appl Environ Microbiol 1985; 49:1314–1322.

12. Mountfort DO, Asher RA. Production of xylanase by the ruminal anaerobic fungus *Neocallimastix frontalis*. Appl Environ Microbiol 1989; 55:1016–1022.

13. Mountfort DO, Asher RA, Bauchop T. Fermentation of cellulose to methane and carbon dioxide by a rumen anaerobic fungus in a triculture with *Methanobrevibacter* sp. strain RAI and *Methanosarcina barkeri*. Appl Environ Microbiol 1982; 44:128–134.

14. Borneman WS, Akin DE, Ljungdahl LG. Fermentation products and plant cell wall–degrading enzymes produced by monocentric and polycentric anaerobic ruminal fungi. Appl Environ Microbiol 1989; 55:1066–1073.

15. Pearce PD, Bauchop T. Glycosidases of the rumen anaerobic fungus *Neocallimastix frontalis* grown on cellulosic substrates. Appl Environ Microbiol 1985; 49:1265–1269.

16. Amos HE, Akin DE. Rumen protozoal degradation of structurally intact forage tissues. Appl Environ Microbiol 1978; 36:513–522.

17. Akin DE, Gordon GLR, Rigsby LL. Comparative fiber degradation by mixed rumen fungi from Australian and U.S.A. cattle. Anim Feed Sci Technol 1989; 23:305–321.

18. Bauchop T, Mountfort DO. Cellulose fermentation by a rumen anaerobic fungus in both the absence and the presence of rumen methanogens. Appl Environ Microbiol 1981; 42:1103–1110.

19. Akin DE, Rigsby LL. Mixed fungal populations and lignocellulosic tissue degradation in the bovine rumen. Appl Environ Microbiol 1987; 53:1987–1995.

20. Akin DE, Chesson A. Lignification as the major factor limiting forage feeding value especially in warm conditions. Proc Intl Grassl Congress 1989; 16:1753–1760.

21. Conchie J, Hay AJ, Lomax JA. Soluble lignin–carbohydrate complexes from sheep rumen fluid: their composition and structural features. Carbohydr Res 1988; 177:127–151.

22. Gaillard BDE, Richards GN. Presence of soluble lignin–carbohydrate complexes in the bovine rumen. Carbohydr Res 1975; 42:135–145.

23. Hartley RD, Akin DE, Himmelsbach DS, Beach DC. Microspectrophotometry of bermudagrass (*Cynodon dactylon*) cell walls in relation to lignification and wall biodegradability. J Sci Food Agric 1990; 50:179–189.

24. Akin DE, Ames-Gottfred N, Hartley RD, et al. Microspectrophotometry of phenolic compounds in bermudagrass cell walls in relation to rumen microbial digestion. Crop Sci 1990; 30:396–401.

25. Spencer RR, Akin DE. Rumen microbial degradation of postassium hydroxide-treated Coastal bermudagrass leaf blades examined by electron microscopy. J Anim Sci 1980; 51:1189–1196.
26. Spencer RR, Akin DE, Rigsby LL. Degradation of potassium hydroxide–treated 'Coastal' bermudagrass stems at two stages of maturity. Agron J 1984; 76:819–824.
27. Akin DE, Willemse MTM, Barton FE II. Histochemical reactions, autofluorescence, and rumen microbial degradation of tissues in untreated and delignified bermudagrass stems. Crop Sci 1985; 25:901–905.
28. Grenet E, Barry P. Colonization of thick walled plant tissues by anaerobic fungi. Anim Fee Sci Technol 1988; 19:25–31.
29. Terrill TH, Windham WR, Hoveland CS, Amos HE. Forage preservation method influences on tannin concentration, intake, and digestibility of sericea lespedeza by sheep. Agron J 1989; 81:435–439.
30. Pritchard GI, Folkins LP, Pigden WJ. The *in vitro* digestibility of whole grasses and their parts at progressive stages of maturity. Can J Plant Sci 1963; 43:79–87.
31. Van Soest PJ. The uniformity and nutritive availability of cellulose. Fed Proc 1973; 32:1804–1808.
32. Borneman WS, Hartley RD, Morrison WH III, et al. Feruloyl and *p*-coumaroyl esterase from anaerobic fungi in relation to plant cell wall degradation. Appl Microbiol Biotechnol 1990; 33:345–351.
33. Akin DE, Benner R. Degradation of polysaccharides and lignin by ruminal bacteria and fungi. Appl Environ Microbiol 1988; 54:1117–1125.
34. Ulyatt MJ, Dellow DW, John A, et al. Contribution of chewing during eating and rumination to the clearance of digesta from the ruminoreticulum. In: Milligan LP, Grovum WL, Dobson A, eds. Control of Digestion and Metabolism in Ruminants. Englewood Cliffs, NJ: Prentice-Hall, 1986:498–515.
35. Welch JG. Physical parameters of fiber affecting passage from the rumen. J Dairy Sci 1986; 69:2750–2754.
36. Wilson JR, Engels FM. Do rumen fungi have a significant direct role in particle size reduction? In: Nolan JV, Leng RA, Demeyer DI, eds. The Rôles of Protozoa and Fungi in Ruminant Digestion. Armidale, Australia: Penambul Books, 1989: 255–257.
37. Akin DE, Lyon CE, Windham WR, Rigsby LL. Physical degradation of lignified stem tissues by ruminal fungi. Appl Environ Microbiol 1989; 55:611–616.
38. Borneman WS, Akin DE. Lignocellulose degradation by rumen fungi and bacteria: ultrastructure and cell wall degrading enzymes. In: Akin DE, Ljungdahl LG, Wilson JR, Harris PJ, eds. Microbial and Plant Opportunities to Improve Lignocellulose Utilization by Ruminants. New York: Elsevier, 1990:325–339.
39. Mackinnon BW, Easton HS, Barry TN, Sedcole JR. The effect of reduced leaf shear strength on the nutritive value of perennial ryegrass. J Agric Sci Camb 1988; 111:469–474.
40. Weston RH, Lindsay JR, Purser DB, et al. Feed intake and digestion responses in sheep to the addition of inorganic sulfur to a herbage diet of low sulfur content. Aust J Agric Res 1988; 39:1107–1119.

41. Akin DE, Borneman WS, Lyon CE. Degradation of leaf blades and stems by monocentric and polycentric isolates of ruminal fungi. Anim Feed Sci Technol 1990; 31:205–221.
42. Akin DE. Histological and physical factors affecting digestibility of forages. Agron J 1989; 81:17–25.

7

Interactions Between the Rumen Chytrid Fungi and Other Microorganisms

ALAN G. WILLIAMS

Hannah Research Institute
Ayr, Scotland

KEITH N. JOBLIN

AgResearch, Grasslands Research Centre
Palmerston North, New Zealand

G. FONTY

Institut National de la Recherche Agronomique
Saint Genes Champanelle, France

I. INTRODUCTION

The ruminant is dependent on the rumen microbial population to predigest ingested feed components and convert them into a utilizable form. In an ecosystem as complex as the rumen, both synergistic and competitive microbial interactions occur (1), and it is well established that metabolic interactions between different populations are required to convert the available substrates into the characteristic products of the rumen fermentation. Interspecies cross-feeding of microbial metabolic products makes available both growth substrates and essential growth factors that increase the complexity of the microbial community, with concomitant effects on the overall course of the fermentation (2,3). The purpose of this chapter is to review the interactions that involve the rumen chytrid fungi and other microbial groups, and to discuss the effects of these interactions on the metabolic activities of the fungi as revealed by recent studies in vitro and in vivo.

The contents of the rumen are heterogeneous, consisting of a solid mass of digesta suspended in a liquid phase. It has been estimated that approximately three-quarters of the ruminal microorganisms are attached to, or closely associated with, the digesta solids (4,5). The particulate material is therefore colonized by a wide array of microbial groups with differing metabolic capabilities (6,7) and is a principal site of microbial activity in the rumen. Diverse groups of microorganisms are thus able to maintain a close spatial association and interact. Separate interacting subpopulations have been recognized by microscopy and through their distinctive enzyme profiles (6,7,8). The chytridiomycete fungi are part of the complex microbial consortium that colonizes plant material in the rumen digesta (Fig. 1), and it is reasonable to assume that interactions with other particle-associated populations occur. In addition, in the rumen, bacteria attach to the fungal sporangia (Fig. 2) (9), and in vitro the fungi form stable associations with methanogenic bacteria (10,11,12). Interactions with hydrogenotrophic (hydrogen-utilizing) species or other metabolic groups have been shown to have significant effects on the biochemical activities of the fungi. The consequences of interspecies interactions of the rumen fungi with various bacterial and protozoal groups are discussed in the following sections.

II. FUNGUS-BACTERIA INTERACTIONS

A. Hydrogenotrophic Bacteria

The rumen chytridiomycete fungi are hydrogen producers (i.e., hydrogenogens) and as such can interact with hydrogenotrophic species. The metabolite

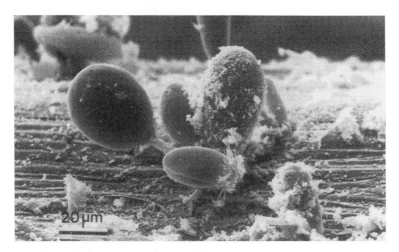

Figure 1 Scanning electron micrograph of fungal sporangia together with bacteria and entodiniomorphid protozoa on ryegrass stem from ruminal digesta. (From K. N. Joblin and D. Hopcroft, unpublished).

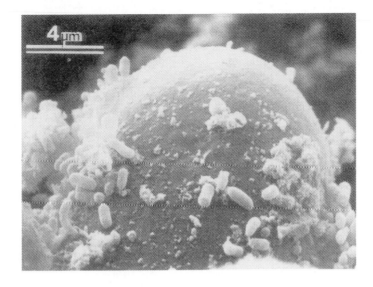

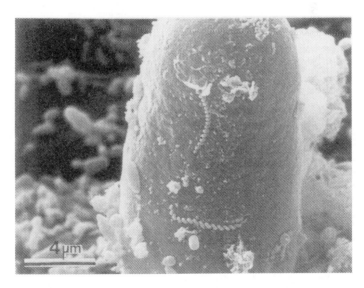

Figure 2 Scanning electron micrographs of bacteria attached to fungal sporangium. (From Ref. 9.)

profile of a hydrogenogen is altered as a direct consequence of the interspecies hydrogen transfer, and in complex communities like the rumen, hydrogenogen–hydrogenotroph interactions are important determinants of the fermentation pattern (2,3).

1. Methanogenic Bacteria

The methanogenic bacteria are the principal hydrogenotrophs in the rumen ecosystem. There is as yet, however, no direct evidence to confirm the existence of a physical association between the fungi and methanogens in the rumen, as has been reported for the rumen ciliate protozoa and methanogenic bacteria. The attachment of methanogens to the external surfaces of some entodinomorphid protozoa has been demonstrated by fluorescence microscopy (13,14,15), and metabolic interactions involving holotrich ciliates and methanogens were subsequently confirmed (16). A physical attachment of methanogens to the fungi may likewise occur, as the identity of the bacteria adhering to the fungal sporangia in situ (Fig. 2) (9) is unknown. However, a close association, rather than direct contact, would be sufficient to enable metabolic interactions to occur. The observation that methanogenic bacteria can be present and persist during the isolation of fungi from rumen contents (10) may be indirect evidence that some form of association does occur in the rumen. The close proximity of the methanogen to the fungal hyphae is evident after coculture in vitro (Fig. 3) (10), but in vivo the rhizoids

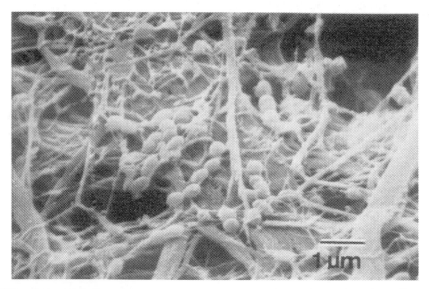

Figure 3   Scanning electron micrograph of an in vitro culture of *N. frontalis* and *Mb. smithii* growing on filter paper. (From Ref. 10.)

are within the plant tissue and possibly less accessible for bacterial contacts (9).

Stable cocultures of methanogenic bacteria have been established in vitro with several strains of the ruminal chytrids *Neocallimastix frontalis, Piromyces* (syn. *Piromonas*) *communis,* and *Sphaeromonas* (syn. *Caecomyces*) *communis* (10,11,12,17–22). Cocultures usually establish readily with both ruminal and nonruminal strains of methanogenic bacteria (i.e., *Methanobrevibacter ruminantium, Mb. smithii,* and *Methanosarcina barkeri*), but there is a report of difficulty in maintaining stable cocultures with a strain of *Mb. ruminantium* (22). The effects of the presence of the methanogen on aspects of the metabolic activities of the fungi are considered in the following sections.

a. Metabolite Formation. The principal metabolites produced during the fermentation of carbohydrates by monocentric and polycentric fungal isolates are acetic acid, lactic acid, formic acid, ethanol, carbon dioxide, and hydrogen; some strains produce traces of succinic acid (23,24,25). When a methanogen is present, however, the fermentation becomes acetogenic; acetic acid production increases, whereas lactic acid and ethanol formation decreases; hydrogen and formic acid do not accumulate (Fig. 4). These effects of cocultivation with a methanogen are common to all three fungal genera examined, irrespective of the carbohydrate growth substrate (Table 1), and are due to hydrogen removal by the hydrogenotrophic methanogen. In monoculture the formation of lactate, ethanol, and hydrogen is essential for the regeneration of reduced nucleotides formed during glycolysis. However, in the presence of a methanogen, the hydrogen formed by the fungus is continuously removed and utilized by the methanogen and does not accumulate. Hydrogen removal by the methanogen results in a low partial pressure of hydrogen, any feedback inhibitory effects of hydrogen on the reoxidation of reduced nucleotides are thus minimized, and the electron carriers are reoxidized during hydrogenesis (Fig. 5). The formation by the fungi of alternative electron-sink metabolites, such as lactate and ethanol, decreases, and in consequence an increased proportion of the hexose-derived pyruvate can be converted into acetate. The potential adenosine triphosphate (ATP) yield is thus increased from 3 to 4 moles per mole of hexose utilized (3). Measurements of fungal biomass indicate that fungal growth is increased in the cocultures (e.g., 19,20,21). Similar metabolic shifts occur as a consequence of interspecies hydrogen transfer between rumen bacterial or protozoal hydrogenogens and methanogenic bacteria (2,16,26). The formic acid formed by the fungus is a precursor of methanogenesis in the rumen (27,28) and can function directly as the interspecies electron carrier (29); formate, therefore, does not accumulate in the fungus–methanogen coculture.

b. Degradation of Plant Cell-Wall Polysaccharides. The rate and extent of cellulose utilization by the rumen fungi *N. frontalis, N. patriciarum, P.*

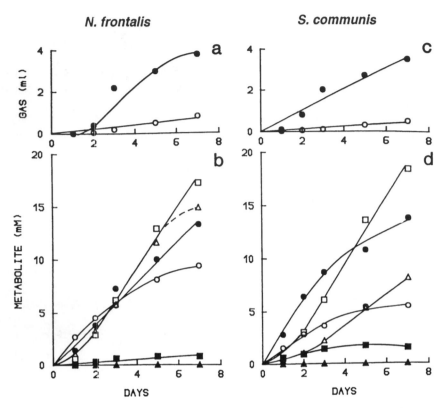

Figure 4   Gas and metabolite formation by *N. frontalis* PNK2 **(a,b)** and *S. communis* B7 **(c,d)** grown in vitro on oat spelt xylan in the absence (open symbols) and presence (closed symbols) of *Mb. smithii*. **(a,c)** Methane (●), hydrogen (○). **(b,d)** Acetic acid (○, ●), formic acid (△, ▲), lactic acid (□, ■).

communis, and *S. communis* are increased in the presence of methanogens (10,12,21,30,31). The effect was first reported by Bauchop and Mountfort (10), who found that a culture of *N. frontalis* utilized 53% of the filter paper substrate in 12 d; however, in coculture with the ruminal methanogen *Methanobrevibacter ruminantium* 83% of the filter paper was used in only 7 days. Cocultures of *N. frontalis* and *Methanosarcina barkeri*, although more effective than the fungal monoculture, degraded less cellulose than the *Mb. ruminantium–N. frontalis* coculture (11). However, a triculture of these three organisms was particularly effective in converting cellulosic substrates to methane (Table 2; 11). The greater cellulose degradation by the triculture was attributed to the methanogen-mediated removal of metabolites that were inhibitory to fungal growth.

Table 1 Metabolite Profiles of Rumen Fungi in the Presence and Absence of Methanogenic Bacteria

Fungus	Methanogen	Substrate	Time (d)	Acetate		Formate		Lactate		Ethanol		Hydrogen		Methane		Ref.
				−	+	−	+	−	+	−	+	−	+	−	+	
N. frontalis	Mb. ruminantium	Cellulose	7	20.0	59.2	21.7	0	12.5	1.0	14.6	8.3	2.5	0	0	6.7	10
P. communis	Mb. ruminantium	Cellulose	8	15.7	39.8	ND		8.9	1.1	7.5	0.7	++	t	0	++	12,21
S. communis	Mb. ruminantium	Cellulose	8	14.3	48.7	ND		5.4	0.7	3.4	0.7	++	t	0	++	12,21
N. frontalis	Mb. smithii	Xylan	8	5.5	17.1	13.8	0	14.7	1.8	++	0	9.4	0	0	25.0	20
P. communis	Mb. smithii	Xylan	7	5.9	13.3	9.8	0	7.0	0.9	++	0	0.4	0	0	3.1	NP
S. communis	Mb. smithii	Xylan	7	4.3	12.6	8.6	0	7.0	0.3	++	0	0.5	0	0	3.1	NP
P. communis	Mb. smithii	Glucose	5	4.7	9.1	10.4	0	ND		ND		0.4	t	0	1.4	18

[a]Concentration mmol/l or ml for gases.

ND = data not presented; ++ = detected but not quantified; t = trace; NP = A. G. Williams and K. N. Joblin, unpublished observations.

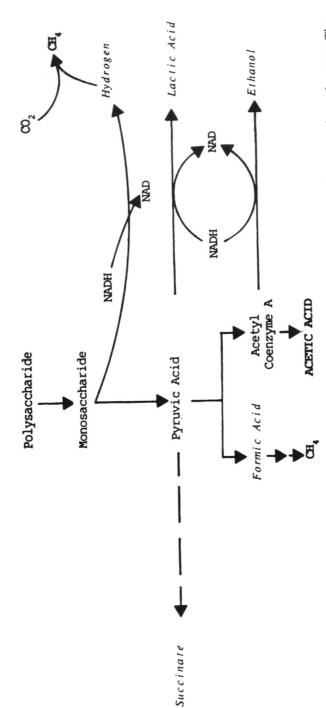

Figure 5   Schematic representation of carbohydrate utilization by rumen fungi in the absence and presence of a methanogen. The products in bold type are the principal products of the fungus–methanogen coculture. Acetate and the products in italics are formed by fungal monocultures.

Table 2 Polysaccharide Utilization by Rumen Chytrid Fungi in the Presence and Absence of a Methanogen

Fungus	Methanogen	Time (d)	Substrate	Utilization (%)			Reference
				Monoculture	Coculture		
N. frontalis	Methanobrevibacter ruminantium	7	Cellulose	42	82		10
N. frontalis	Methanobrevibacter sp.	10	Cellulose	46	84		11
N. frontalis	Methanosarcina barkeri	10	Cellulose	46	65		11
N. frontalis	Methanosarcina barkeri + Methanobrevibacter sp.	10	Cellulose	46	94		11
N. frontalis RE1	Methanobacterium arboriphilus	14	Cellulose	89	93		30
N. frontalis RE1	Methanobacterium bryantii	14	Cellulose	89	93		30
N. frontalis RE1	Methanobrevibacter smithii	14	Cellulose	89	96		30
N. patriciarum	Methanobacterium arboriphilus	14	Cellulose	85	94		30
N. patriciarum	Methanobacterium bryantii	14	Cellulose	85	93		30
N. patriciarum	Methanobrevibacter smithii	14	Cellulose	85	95		30
P. communis	Methanobrevibacter ruminantium	8	Cellulose	50	75		12
P. communis	Methanobacterium arboriphilus	14	Cellulose	80	95		30
P. communis	Methanobacterium bryantii	14	Cellulose	80	94		30
P. communis	Methanobrevibacter smithii	14	Cellulose	80	93		30
S. communis	Methanobrevibacter ruminantium	8	Cellulose	12	30		12
S. communis FG10	Methanobacterium arboriphilus	14	Cellulose	79	94		30
S. communis FG10	Methanobacterium bryantii	14	Cellulose	79	94		30
S. communis FG10	Methanobrevibacter smithii	14	Cellulose	79	96		30
N. frontalis	Methanobrevibacter smithii	7	Xylan	81	89		20
S. communis	Methanobrevibacter smithii	7	Xylan	52	81		20
P. communis	Methanobrevibacter smithii	7	Xylan	83	87		20

The rate and extent of hemicellulose utilization and degradation by rumen fungi are also improved by cocultivation with a methanogen (Table 2; 20). Although *N. frontalis, S. communis,* and *P. communis* are able to extensively degrade oat spelt arabinoxylan (≥90%), the rate and extent of utilization of this hemicellulosic polysaccharide were increased in the presence of *Mb. smithii* (Fig. 6). The *N. frontalis* monoculture was also less effective than the coculture with *Mb. smithii* after prolonged incubation periods of 18 d. Within 5 d, 82% of the xylan had been utilized by the coculture as compared with 59% by the fungal monoculture; after 18 d the utilization values had increased to 84% and 78%, respectively (20). Growth of fungus–methanogen associations on xylan results in acetogenic fermentations similar to those found for growth on cellulose (Table 1). The rate of polysaccharide utilization and the proportions of metabolites formed during rumen bacterial cellulolysis and hemicellulolysis are also affected by methanogenic bacteria (26,32,33).

The presence of methanogens in in vitro fungal cultures often results in enhanced degradation of intact plant tissue, although the effects are less pronounced than with paper or isolated hemicellulosic substrates. It is probable that the architecture of the plant cell-wall reduces substrate accessibility.

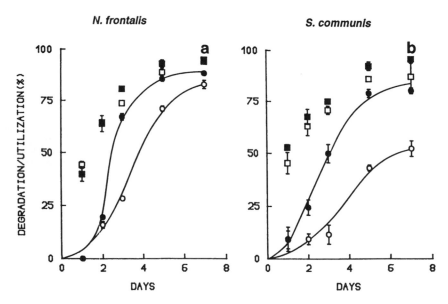

Figure 6 Degradation (□, ■) and utilization (○, ●) of oat spelt xylan by *N. frontalis* PNK2 **(a)** and *S. communis* B7 **(b)** when grown in the presence (closed symbols) or absence (open symbols) of the methanogen *Mb. smithii.*

The presence of a methanogen may reduce the extent of the lag period before digestion begins. The extent of the enhancement of cell-wall degradation by cocultures depends on the strain of fungus and the species of methanogen involved. Digestion of ammoniated barley straw by *N. frontalis* PNK2, *N. frontalis* RK21, and *P. communis* RK11 was improved by 35%, 22%, and 14%, respectively, in coculture with *Mb. smithii* (22). In contrast, in coculture with *Ms. barkeri* digestion of similar tissue by *N. frontalis* PNK2 increased by only 19%, while that of the other fungal strains was not significantly improved (K. N. Joblin and C. S. Stewart, unpublished). When two strains of *N. frontalis* were incubated with the methanogen *Mb. ruminantium*, straw digestion by strain PNK2 increased but strain RK21 was not affected (17). The reasons for strain differences in the effectiveness of the interactions have not been established.

The extent of lucerne stem degradation in vitro was slightly increased in the initial stages of the fermentation by cocultivation of *N. frontalis* PNK2 and *P. communis* B19 with *Mb. smithii* (34). The principal component sugars were removed simultaneously from the intact stem material by both the monoculture and the coculture. In the presence of *Mb. smithii* the extent of glucose, xylose, and arabinose removal from plant tissue by *P. communis* was increased in the initial stages of the fermentation (Fig. 7). Analysis of

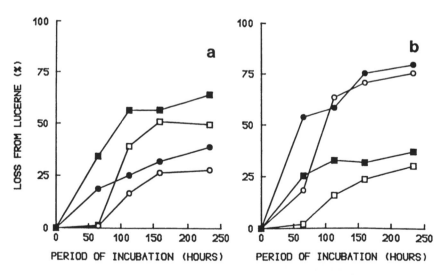

Figure 7 Loss of dry matter and plant cell-wall component monosaccharides from lucerne stem during in vitro incubation with *P. communis* B19 in the presence (closed symbols) and absence (open symbols) of *Mb. smithii*. (a) Dry matter (○, ●), glucose (□, ■). (b) Arabinose (○, ●), xylose (□, ■). The residues from triplicated incubators were combined before analysis.

the digested stem residue indicated that cellulose was removed more rapidly than the xylan component, and of the hemicellulosic monomers arabinose was solubilized more rapidly than xylose (34). The presence of a methanogen in in vitro cultures of the fungi reduces the lag period before digestion of lignocellulose begins but appears to have little stimulatory effect on the rate of digestion (Fig. 7) (22,34).

The enhancement of fungal utilization of plant cell-wall polysaccharides is of particular significance in the rumen ecosystem, where the fungi are believed to have an important role in fiber digestion (see Section IV). The enhanced polysaccharide utilization by the coculture arises, in all probability, from several consequences of the interaction. The higher ATP yield of the fermentation enables increased fungal growth to take place, and the higher utilization may merely be a consequence of the increased fungal biomass. Cocultivation may also prevent the build-up of inhibitory metabolites. Decreased lactate accumulation may be beneficial, as lactic acid has been shown to have an adverse effect on rumen protozoal metabolism (35) and ruminal cellulolysis (36) and may likewise inhibit fungal activity. It has been suggested that formate, lactate, and ethanol might inhibit in vitro cellulolysis by *N. frontalis* (11,37). The association between methanogens and fungi not only influences fungal metabolite formation but also has a pronounced effect on the polysaccharide-degrading enzymes of the fungi. These effects are considered in the following section.

   c. Formation of Polysaccharide-Degrading Enzymes. The rumen chytrid fungi form a wide range of plant cell-wall–degrading enzymes (38–42). The first indication that coculturing fungi with methanogens affected enzyme activities came from studies on fungal cellulolytic enzymes (38). The activity (U/ml) and yield (U/g cellulose degraded) of the β-glucosidase, carboxymethylcellulase, and cello-oligosaccharidase of *N. frontalis* PN1 were increased by cocultivation with *Methanospirillum hungatei*. The cotton-solubilizing activity of the extracellular cellulase of *N. frontalis* RK21 was markedly increased by cocultivation with methanogenic bacteria (43). However, it is not only cellulose-degrading enzymes that are affected by cocultivation, as Joblin et al. (20) demonstrated that the specific activities of extracellular hemicellulose-degrading enzymes formed by *N. frontalis, P. communis,* and *S. communis* were increased by cocultivation with *Mb. smithii* on monosaccharides, hemicellulosic polymers, and plant material (Table 3). The specific activities of both the extracellular and cell-associated enzymes were consistently higher in an *N. frontalis–M. smithii* coculture, as compared with the activities in the *N. frontalis* monoculture, when the fungus was cultured on xylan (Table 4). As protein secretion also increased in the coculture, the enzyme titers (U/ml) as well as the specific activities (U/mg protein) increased. The xylanolytic activity of *Neocallimastix* isolates cocultured with

Table 3 Extracellular Enzyme Activities[a] of Rumen Fungi Grown With and Without *Mb. smithii*

Fungus	Carbon source	α-L-Arabinofuranosidase		β-D-Xylosidase		β-D-Glucosidase		Xylanase	
		−	+	−	+	−	+	−	+
N. frontalis PNK2	Glucose	7.1	30.8	40.7	298	40.3	149	128	739
	Xylan	14.2	16.2	140	297	27.8	159	111	459
	Lucerne	7.8	38.7	58.8	335	33.5	134	343	913
S. communis B7	Glucose	1.2	2.4	63.0	333	26.6	123	134	247
	Xylan[b]	4.1	23.1	26.4	239	NA	NA	37.6	145
	Lucerne	2.0	1.5	11.0	39.4	7.3	27.4	118	195
P. communis B19	Xylose	21.9	33.8	134	606	4.8	19.8	294	412
	Xylan[b]	56.9	80.5	74.8	114	104	239	193	315
	Lucerne	7.8	21.9	73.1	229	21.8	76.1	353	631

[a]Specific activities mU/mg protein. One unit of enzyme activity released 1 μmol of product in 1 min. Cultures were grown for 5 d or [b]7 d.
NA = not assayed.
Source: Data from Refs. 20, 34; A. G. Williams and K. N. Joblin, unpublished.

Table 4 Specific Activities[a] of Cell-Associated and Extracellular Fibrolytic Enzymes Formed by *N. frontalis* Grown on Xylan in the Absence of Presence of *Mb. smithii*

Enzyme	Location	Period 1[b] −	Period 1[b] +	Period 2[c] −	Period 2[c] +
α-D-Galactosidase	C	1.1	4.2	0.8	4.6
	E	ND	3.5	0.6	5.3
β-D-Galactosidase	C	1.4	6.9	1.2	9.1
	E	0.7	6.7	1.4	9.9
α-D-Glucosidase	C	2.5	2.9	3.0	3.1
	E	0.9	3.2	2.6	3.9
β-D-Glucosidase	C	49.0	120	45.9	171
	E	31.5	155	62.3	217
β-D-Cellobiosidase	C	7.3	27.8	7.9	32.6
	E	2.6	31.2	8.2	46.6
α-L-Arabinofuranoside	C	41.2	42.4	9.0	19.5
	E	16.9	21.4	36.6	47.1
β-D-Xylosidase	C	435	385	71.1	144
	E	163	318	317	513
Xylanase	C	463	630	297	806
	E	362	1318	663	1703
Hemicellulase	C	462	575	353	695
	E	427	1145	585	1814
Carboxymethylcellulase	C	116	180	110	324
	E	112	332	150	365
Cellulase	C	12.6	36.7	8.7	36.2

[a]Specific activities (mU/mg protein) are means of measurements made on [b]days 4, 5, and 8 (Period 1) and [c]days 12, 15, and 18.
C = cell associated; E = extracellular; ND = not detected.
Source: Data adapted from Ref. 20.

Methanobacterium formicicum is also increased (31). It is now apparent that the effects of the synergy affect a range of polysaccharide-degrading enzymes and extend beyond those originally described for cellulolysis. The higher enzyme activities in cocultures are no doubt a factor in explaining the enhanced in vitro polysaccharide and plant degradation by fungus–methanogen cultures.

The higher activities may be attributable to increased fungal biomass or removal of potentially inhibitory metabolites in the cocultures, as discussed earlier. Differences in fungal growth rate and sugar accumulation between monocultures and cocultures may also influence enzyme activity. Fungal growth rate could influence the activities of any enzymes that are growth-dependent. The activities of polysaccharolytic enzymes formed by some fibrolytic rumen bacteria are dependent on both the stage and the rate of growth of the microorganism (44).

The level of free sugars, however, would seem to be a key factor affecting the level of fungal enzyme activities (20). Soluble sugars accumulate during fungal degradation of complex polysaccharides and plant material (11,20, 34,42,45,46), although with the exception of arabinose (20,46) they are subsequently utilized. The extent and period of sugar accumulation during fungal growth on xylan was markedly reduced in fungus–methanogen cocultures (20), and it has been reported that sugars did not accumulate in cocultures on cellulose (11). Polysaccharide-degrading enzymes formed by the rumen fungi are produced constitutively, and although their activities are not inhibited by low levels of free sugars, their formation is subject to catabolite repression (38,39,40,42,45,46,47). Fungi utilize soluble carbohydrates preferentially, and polysaccharide degradation does not take place until the soluble sugars have been consumed (e.g., 24,48). The addition of soluble sugars to cultures growing on a polysaccharidic carbon source suppresses the production of the polysaccharide-degrading enzymes (38,42,45). In cocultures, the more rapid utilization of the released sugars results in a much-reduced or nil exposure of fungi to free sugars, and catabolite repression on enzyme formation is lessened. As a consequence, polysaccharide depolymerase and glycohydrolase activities are higher in cocultures. Endopeptidase (i.e., aminoacylamidase) activity of *N. frontalis* PNK2, however, was not increased by cocultivation of the fungus with *Mb. smithii* (20).

d. Ionophore Resistance. The carboxylic ionophore monensin depresses growth and inhibits the fibrolytic activity and carbohydrate metabolism of axenically grown cultures of *N. frontalis* and *P. communis* (19,49,50,51). Another ionophore, lasalocid, also depresses growth, glucose uptake, and metabolite formation (19). However, in the presence of *Mb. smithii* the susceptibility of both fungi to the effects of monensin and lasalocid was reduced (19). The mechanism by which the methanogens enhance the tolerance of the fungi to the antibiotics has yet to be established. Monensin and lasalocid catalyze an exchange of cations and protons across cell membranes, with a concommitant perturbation of the protonmotive force. The higher energy yield of the fungus in the presence of the methanogen may ensure that adequate ATP is available to restore the protonmotive force and offset the deleterious effects of the ionophores (19,52).

2. Other Hydrogenotrophic Species

Although the hydrogenotroph–fungus interactions studied in detail have involved methanogenic species, other hydrogenotrophic bacteria are present in the rumen and may therefore interact with ruminal chytrid fungi. In the rumen ecosystem, methanogenesis is the principal mechanism for the disposal of electrons and maintenance of a low hydrogen partial pressure (53). However, other potential electron acceptors that occur in this intestinal environment are sulfate, nitrate, carbon dioxide, oxygen, and fumarate (2). Sulfate-reducing bacteria are present in the rumen (54), and some ruminal bacteria can use nitrate as an electron acceptor (55). *Wolinella succinogenes* derives the energy necessary for growth from fumarate reduction (56), whereas some strains of *Selenomonas ruminantium* utilize extracellular hydrogen for propionate formation (57). Other ruminal hydrogenotrophs are able to reduce carbon dioxide to acetic acid. The acetogenic species *Eubacterium limosum* and *Acetitomaculum ruminis* participate in the utilization of hydrogen in the rumen (58,59,60). Syntrophic associations involving hydrogenotrophs, other than methanogens, have been demonstrated to occur in nonruminal environments (2), but as yet information is very limited on the occurrence and effects of interspecies hydrogen transfer between hydrogenogens and nonmethanogenic hydrogenotrophs in the rumen (3).

It has been established, however, that the chytrid fungi interact with nonmethanogenic hydrogenotrophs. Richardson and Stewart (61) demonstrated that lactate formation by *N. frontalis* did not occur in coculture with the hydrogenotroph *Selenomonas ruminantium*. The principal metabolites of cellulose fermentation by *N. frontalis* after cocultivation with lactate-utilizing *Sel. ruminantium* strains were acetate, formate, propionate, and carbon dioxide; succinate was not detected, and hydrogen did not accumulate. Lactate production was reduced by 80% following cocultivation with a nonlactilytic strain of *Sel. ruminantium*, confirming that interspecies hydrogen transfer had taken place. The bacterial utilization of hydrogen obviated the need for lactate formation as an electron-sink product by the fungus (61). The principal products of the associative fermentation of *S. communis* or *P. communis* and *Selenomonas ruminantium* WPL were propionate and acetate (21). However, some lactate and ethanol accumulated during these incubations.

Cocultures of rumen fungi have also been established with the acetogenic hydrogenotroph *Eubacterium limosum* (strain DSM 20543). Filter paper degradation by *N. frontalis* MCH3 and *P. communis* FL was not affected by cocultivation with *E. limosum*, whereas cellulolysis by *S. communis* was decreased (Table 5). Interspecies hydrogen transfer resulted in increased acetate accumulation although the formation of both formate and lactate decreased (Table 6). Acetate production per mole of hexose fermented was increased by approximately 50% in the *E. limosum*–fungus coculture (A. Bernalier and G. Fonty, unpublished data).

B. Metabolite-Utilizing Bacteria

The utilization of the catabolic products and metabolites of one group of organisms by other groups is necessary to sustain a diverse microbial popula-

Table 5 Effect of Cocultivation of Rumen Fungi and Nonmethanogenic Bacteria on Cellulose and Xylan Utilization

Fungus	Bacterial species	Bacterial type	Effect of cocultivation[a] Cellulolysis	Reference
N. frontalis RE1	Selenomonas ruminantium JW2	L	–	61
N. frontalis RE1	Sel. ruminantium JW13	L	–	61
P. communis FL	Sel. ruminantium WPL	L	–	21
S. communis FG10	Sel. ruminantium WPL	L	+	21
N. frontalis PNK2	Megasphaera elsdenii J1/M	L	+	17,18
N. frontalis MCH3	M. elsdenii DSM 20460	L	–	NP[b]
P. communis FL	M. elsdenii DSM 20460	L	–	NP[b]
S. communis FG10	M. elsdenii DSM 20460	L	–	NP[b]
N. frontalis	Veillonella parvula (syn. alcalescens)	L	+	18
N. frontalis PNK2	V. parvula S113, L59	L	+	17
N. frontalis	Ruminococcus albus	C	–	17,18,71
N. frontalis	R. flavefaciens	C	–	17,18,68,71
P. communis FL	R. flavefaciens	C	–	68
S. communis FG10	R. flavefaciens	C	+	68
N. frontalis	Fibrobacter succinogens	C	+	18
N. frontalis MCH3	Fibrobacter succinogenes	C	0	68
P. communis FL	Fibrobacter succinogenes	C	0	68
S. communis FG10	Fibrobacter succinogenes	C	0	68

Table 5 Continued

Fungus	Bacterial species	Bacterial type	Effect of cocultivation[a]	Reference
N. frontalis MCH3	Eubacterium limosum	H	0	NP[b]
P. communis FL	E. limosum	H	0	NP[b]
S. communis	E. limosum	H	–	NP[b]
			Xylanolysis	
N. frontalis PNK2	V. parvula (syn. alcalescens)	L	±	65
N. frontalis PNK2	Sel. ruminantium lactilytica	L	++	65
N. frontalis PNK2	Sel. ruminantium ruminantium	L–	++	65
P. communis B19	Sel. ruminantium ruminantium	L–	+	NP[c]
N. frontalis PNK2	Butyrivibrio fibrisolvens NCD0 2249	X	+	65
N. frontalisPNK2	Bacteroides (syn. Prevotella) ruminicola ruminicola	X	++	65
P. communis B19	Bact. ruminicola ruminicola	X	+	NP[c]
N. frontalis PNK2	R. flavefaciens FD1	C	±	65
N. frontalis PNK2	Succinivibrio dextrinosolvens	S	++	65

[a]Polysaccharide utilization in coculture as compared with monocultures; 0, no effect; –, utilization inhibited; ±, small increase; +, utilization increased to above values for either monoculture; + +, synergy (i.e., coculture utilization > sum of utilization of bacterial and fungal monoculture values).

[b]A. Bernalier and G. Fonty, unpublished results.

[c]A. G. Williams and K. N. Joblin, unpublished results.

L– = lactate not utilized; L = lactate utilizer; C = cellulolytic; H = hydrogenotrophic; X = xylanolytic; S = saccharolytic.

Table 6 Cellulose Fermentation Products of *N. frontalis* Grown in Axenic Culture or in Association with *Eubacterium limosum*[a]

Culture	Metabolite formation[b]					
	Hydrogen	Formic acid	Acetic acid	Butyric acid	Lactic acid	Ethanol
N. frontalis MCH3	46.5	65.4	39.3	0	42.1	5.4
E. limosum 20543	0	0	99.6	49.4	0	0
N. frontalis MCH3/ *E. limosum* 20543	35.4	13.7	67.1	17.0	23.2	4.5

[a]The substrate for *E. limosum* was glucose.
[b]mol product/100 mol hexose fermented.
Source: Data from Ref. 79; A. Bernalier and G. Fonty, unpublished.

tion. Nutrient cross-feeding interactions support the growth and survival of species that are not primary degraders in the ecosystem. Bacterial nutritional interdependence is exemplified by the cross-feeding on polysaccharide hydrolysis products and the metabolites formed (2,62). The rumen fungi also contribute to the interspecies transfer of nutrients, as they are actively fibrolytic and their extracellular enzymes release free sugars into the environment (20,46). The released component sugars will be utilized competitively by the primary degrader and nonfibrolytic saccharolytic species. Interspecies cross-feeding will occur for sugars that are released by fungal action but are not subsequently utilized (e.g., arabinose). In addition, the metabolites formed by the fungi can be metabolized by other microorganisms in the rumen. The fungi themselves are dependent on nutritional cross-feeding to meet their growth requirements. Growth factors that are required by the fungi have been shown to include B vitamins and heme; other compounds (e.g., branched-chain volatile fatty acids, amino acids) also stimulate growth (9).

The fungal metabolites (Fig. 4), with the exception of acetate, can all be utilized and metabolized further by biochemical mechanisms that are known to be present in ruminal microorganisms and to operate in the rumen ecosystem (2,3,51,61). Some of the potential pathways that could be involved are shown in Fig. 8. The consequences of metabolic interactions, other than interspecies hydrogen transfer, on the polysaccharolytic activity of the rumen fungi are considered in the following sections and summarized in Table 5.

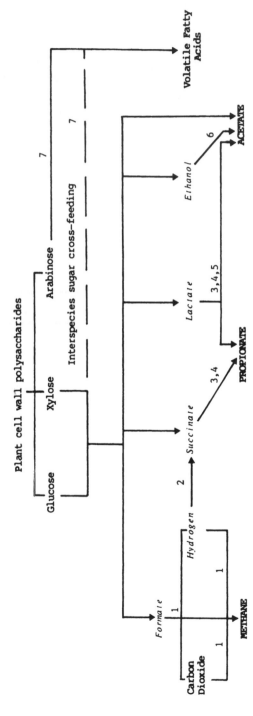

Figure 8 Schematic representation of potential interspecies cross-feeding interactions involving fungal metabolites. The metabolites shown in italics can be further metabolized to the products shown in bold type. The microorganism involved in the interconversions are: 1, methanogens; 2, *Wolinella succinogenes*; 3, *Selenomonas* spp.; 4, *Veillonella* spp.; 5, *Megasphaera elsdenii*; 6, *Clostridium* spp.; 7, saccharolytic species. (Adapted from Ref. 61.)

1. Lactate-Utilizing Bacteria

Rumen fungi have been cocultured with lactate utilizers from three different bacterial genera. The effects on polysaccharide utilization have, however, not been consistent. Some *Selenomonas* spp., in addition to consuming hydrogen, are able to ferment soluble sugars and lactate and are also able to utilize succinate produced by other microorganisms (54,63,64). In fungus-selenomonad cocultures the principal metabolites formed were acetate, formate, propionate, and carbon dioxide. Although hydrogen and lactate formation were considerably reduced, Richardson and Stewart (61) concluded that interspecies hydrogen transfer exceeded lactate cross-feeding; succinate was not detected in the cocultures.

The rate of cellulose breakdown by *N. frontalis* RE1 was decreased by *Sel. ruminantium lactilytica* strain JW13 and *Sel. ruminantium* subsp. *ruminantium* JW2 (61). The cellulolytic activity of *P. communis* was likewise depressed in the presence of a selenomonad (*Sel. ruminantium* strain WPL) (21). However, the breakdown of filter paper by *S. communis* was increased by cocultivation with strain WPL (21); some strains of *Sel. ruminantium* have also been shown to enhance cellulolysis by *N. frontalis* PNK2 (17). The reasons for these strain differences have not been established, but some of the strains may have a higher substrate affinity (i.e., lower K_m) for the released glucose. Such strains would be able to compete more effectively for the available substrate than the fungi and would dominate the coculture.

Fungal utilization of xylan is also increased by cocultivation with *Sel. ruminantium* (Table 5). The interaction of both subspecies of *Sel. ruminantium* with *N. frontalis* PNK2 was synergistic, as after 48 h the utilization (70–75%) by the coculture exceeded the sum of the individual bacterial and fungal monoculture values (approximately 15% and 35%, respectively) (65). The more efficient fermentation of the released sugars by the coculture would minimize catabolite repression of degradative enzymes, enabling the fermentation to proceed more rapidly.

Cocultivation with other lactate-utilizing strains can also increase fungal cellulolysis and hemicellulolysis. *Megasphaera elsdenii* has an important role in ruminal lactate utilization (54,66), and in coculture *M. elsdenii* strain J1/M enhanced filter-paper solubilization by *N. frontalis* PNK2 (17,18). However, Bernalier and Fonty (unpublished results) have noted a marked decrease in filter-paper degradation by *N. frontalis* MCH3, *P. communis* FL, and *S. communis* when these fungi were cocultured with *M. elsdenii* strain DSM 20460. There are also strain differences in the effect of *Veillonella* species. *Veillonella parvula* (syn *V. alcalescens*) strains L59 and S113 improved the degradation of filter paper by *N. frontalis* (17,18). Not all strains of *V. parvula* are effective in coculture, and the effects of *V. parvula* on hemicellulolysis by *N. frontalis* PNK2 were inconclusive. After 48 h no effect was observed (65); however, after 63 h the utilization by the coculture was 61%, as compared with 54% by the fungal monoculture (A. G. Williams and K. N. Joblin, unpublished). The involvement of *V. parvula* strains in interspecies metabolite cross-feeding may, however, be of some significance, as in addition to fermenting lactic acid, the microorganism is able to decarboxylate succinic acid to propionic acid (54).

2. Fibrolytic Bacteria

Cocultures of *N. frontalis* or *P. communis* with either *Ruminococcus albus* or *R. flavefaciens* were less effective than monocultures of the fungus in digesting xylan (65), filter paper (21,67,68), barley straw (17,18), maize stem (69,70), and wheat straw (70). The bacterial antagonistic effect of the ruminococci to fungal cellulolysis was recovered in the culture supernatant of *R. albus* and *R. flavefaciens* and was associated with proteins of differing molecular size (71). However, straw digestion was improved in cocultures of *N. frontalis* and *Fibrobacter* (syn *Bacteroides*) *succinogenes* (18). Electron-microscopic examination of some cocultures containing ruminococci indicated extensive bacterial colonization of the straw. It has been suggested that in some circumstances the initial high bacterial growth rates may result in the exclusion of the fungal zoospores from the attachment sites, thereby preventing their germination and the subsequent normal development of the fungus (17). Delayed fungal establishment may have occurred also in fungus–*R. flavefaciens* cocultures growing on xylan, as a time-dependent improvement in xylanolysis was observed (65).

The utilization of xylan by *N. frontalis* PNK2 apparently increased during cocultivation with *Butyrivibrio fibrisolvens* NCDO 2249, but the increase was probably due to the bacterial xylanolysis (65). However, the interaction with a hemicellulolytic strain of *Prevotella* (syn. *Bacteroides*) *ruminicola* subsp. *ruminicola* was synergistic in that utilization by the coculture was approximately 87% after 48 h incubation. The bacterial and fungal monocultures utilized only 22% and 35%, respectively (65). Xylanolysis by *P. communis* B19 was also increased in coculture with *Bact. ruminicola* (Table 5).

3. Saccharolytic Bacteria

Succinivibrio dextrinosolvens is able to utilize a range of soluble carbohydrates (54) and forms a wide range of glycoside hydrolase enzymes that could enable products of polysaccharide breakdown to be utilized (72). In cocultivation with *N. frontalis* PNK2, hemicellulose breakdown by the fungus in a 48-h incubation was increased from 35% to approximately 75% (65). Hemicellulose degradation products released by the fungal enzymes were utilized by *Suc. dextrinosolvens* and therefore did not accumulate and repress fungal activity. However, in cocultures in which the bacterial partner is unable to utilize the released pentoses, sugars can accumulate and xylan utilization may be inhibited to some extent. Xylan utilization by *N. frontalis* PNK2 was inhibited, for example, when cultured in association with nonsaccharolytic strains of *Streptococcus bovis* and *Lachnospira multiparus* (65).

4. Proteolytic Bacteria

N. frontalis strains and other classes of rumen fungi have been shown to produce extracellular proteinases (73). There are no reports of synergism or competition between bacteria and fungi during proteolysis, but the proteolytic activity of a defined mixture of six rumen bacteria growing on a solid substrate was found to increase after incubation in vitro with *N. frontalis* PNK2 (74), presumably because of the contribution from fungal proteinases. Endopeptidase activity of *N. frontalis* PNK2 was not affected by cocultivation with the methanogen *Mb. smithii* (20).

III. FUNGUS–PROTOZOA INTERACTIONS

Many different species and genera of ciliate protozoa coexist with bacteria and fungi in the rumen (75), and both in vivo and in vitro studies have established that protozoa ingest bacteria and strongly influence bacterial populations (76,77). It is probable that protozoa also influence the fungal population in the rumen (78,79).

Fungal zoospores are small (80) and of similar size to bacteria such as *Selenomonas ruminantium* that are taken up and digested by protozoa (77), so it might be expected that protozoa would ingest zoospores. At present there is no direct evidence for the protozoal predation of zoospores, but it has been noted that the zoospore population density in the rumen increased severalfold when protozoa were removed (81). In ruminal digesta samples, protozoa have been observed at sites of mature fungal sporangia (Fig. 1) (9,82), suggesting that protozoa may be attracted to such sites (82) and could feed on zoospores released from sporangia.

Ciliate protozoa attach to plant particles (75,83) and have an important part either directly or indirectly in ruminal fiber digestion (83). Fungi also attach to plant fragments and play a role in fiber digestion (9,82), and both fungi and protozoa are attracted to sites of tissue damage (84). The only information on possible protozoa–fungi interactions during fiber degradation comes from in vitro studies using purified plant polysaccharides. In one such investigation it has been noted that although the holotrich *Dasytricha ruminantium* had no effect on cellulolysis by *N. frontalis*, a mixed protozoal population containing medium-sized entodiniomorphs markedly reduced fungal cellulolysis (79,82). In the case of xylanolysis, interactions between fungi and protozoa have been assessed (79; A. G. Williams and K. N. Joblin, unpublished data). Mixed holotrichs or *Polyplastron multivesiculatum* had no inhibitory or synergistic effects on xylan degradation by *N. frontalis*, whereas there was evidence to indicate that xylan degradation was increased when the fungus was coincubated with either *Eudiplodinium maggii* or *Entodinium simplex* (Table 7). The increased xylan degradation noted in this study may be due to a contribution from protozoal fibrolytic enzymes.

Predation of protozoa by other protozoa is known to occur (75,85), and in vitro studies have indicated that protozoa may also prey on fungi. When protozoa of different genera were added to *P. communis* growing on plant fragments (K. N. Joblin and A. G. Williams, unpublished data), scanning electron microscopy revealed that *P. multivesiculatum, Eu. maggii,* and *Entodinium* spp. ingested fungal rhizoids (Fig. 9a,b) and in some cases fungal sporangia (Fig. 9c,d and Fig. 10). No evidence for engulfment by *Ophryoscolex caudatus* or by the holotrichs *Isotricha* spp. and *Dasytricha ruminantium* was found, although these protozoa sometimes became lodged among fungal rhizoids. In a parallel experiment, these workers assessed the capacity of protozoa to digest fungal tissue (K. N. Joblin and A. G. Williams, unpublished data). When *Eu. maggii, P. multivesiculatum, Entodinium* spp., and mixed protozoa were incubated with ^{14}C-labeled *P. communis*, a significant increase in ^{14}C solubilization occurred (Table 8). Although more work is needed to clarify these observations and to extend work to the in vivo situation, it can be concluded that protozoa have the capacity to affect the development of fungi in the rumen. Jouany (86) concluded that the number of zoospores present was not affected by protozoa, but other studies have indicated that the zoospore numbers may increase or decrease after defaunation (81,87,88,89). Further studies are thus needed to elucidate the effects of the ciliate protozoa on the rumen fungal population.

Table 7   Effect of Protozoa on Xylan Degradation by *N. frontalis*

	Degradation (%) after 24 h	
Inoculum	P[a]	S[b]
N. frontalis	73.9 ± 1.9	63.9 ± 4.6
+ Holotrichs	78.5 ± 2.8	61.1 ± 2.9
+ *P. multivesiculatum*	77.9 ± 2.2	66.7 ± 2.8
+ *Eu. maggii*	75.3 ± 1.8	71.5 ± 0.9
+ *E. simplex*	80.4 ± 0.6	64.8 ± 6.3

[a]Protozoa added to *N. frontalis* PNK2 preestablished for 48 h.
[b]Simultaneous inoculation of fungus and protozoa.
Source: Ref. 79.

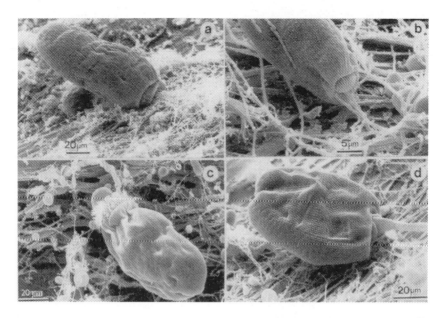

Figure 9 Scanning electron micrographs of protozoa incubated for 5 h with *P. communis* growing on straw. **(a)** *Polyplastron multivesiculatum* and **(b)** an *Entodinium* ingesting rhizoids; **(c)** an *Entodinium* attached to a sporangium; **(d)** an *Entodinium* that had engulfed a sporangium.

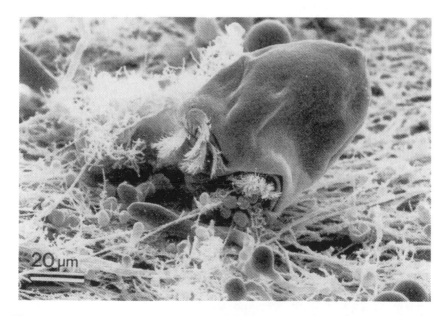

Figure 10 Scanning electron micrograph showing *Eudiplodinium maggii* ingesting immature sporangia after 5 h coincubation with *P. communis* cultured on straw.

Table 8   Release of ^{14}C During in Vitro Incubation of ^{14}C-Labeled
P. communis with Protozoa

Protozoal inoculum	^{14}C Released[a]	
	2 h	18 h
Control (no protozoa)	1030 ± 191	2017 ± 39
Eudiplodinium maggii	932 ± 324	6046 ± 230
Polyplastron multivesiculatum	1440 ± 241	6335 ± 1750
Mixed protozoa (A type)	1098 ± 142	6135 ± 1621
Entodinium spp.	1390 ± 327	8452 ± 859

[a]Mean and standard deviation (d.p.m.) from triplicate cultures.
Source: K. N. Joblin and A. G. Williams, unpublished data.

IV. FUNGI–BACTERIA INTERACTIONS IN VIVO

The effect of fungi on plant cell-wall degradation has been assessed in the developing rumen. Young lambs were placed in sterile isolators 24 h after birth, before the natural establishment of cellulolytic microorganisms (bacteria, fungi, and protozoa), and were reared gnotobiotically according to the method described by Fonty et al. (90). The lambs were inoculated with fungi as the sole cellulolytic microorganisms, and the ability of the established microbial population to digest wheat straw and ryegrass hay was measured in sacco. *N. frontalis* MCH3 removed 29.6% and 32.7% of the dry matter from wheat straw after 48 h and 72 h incubation in nylon bags, respectively; the corresponding cellulose (Van Soest) digestibility was 21.2% and 24.4% (91). After 48 h incubation the *N. frontalis*-containing population degraded 71% of dry matter and 36% of cellulose from ryegrass hay. The efficiency of *P. communis* FL in degrading these two substrates was similar to that of *N. frontalis* (G. Fonty, F. Bonnemoy, and P. Gouet, unpublished data). Dry matter and cellulose disappearance was higher in conventionally reared lambs and in isolated lambs harboring *Fibrobacter succinogenes* S85 or *R. flavefaciens* 007 as the sole cellulolytic microorganisms but was similar in lambs inoculated with *R. albus* (90). However, scanning electron microscopy confirmed that the degradation of maize and lucerne stem fragments was similar in animals inoculated with fungi and in conventionally reared lambs (92). In both cases the fungi preferentially colonized lignified tissues.

A. Fungi–Cellulolytic Bacteria Interactions

Interrelationships between fungi and cellulolytic bacteria have been investigated using two animal models (G. Fonty, A. G. Williams, K. J. Cheng, F. Bonnemoy and P. Gouet, unpublished data). Interactions between fungi and

Table 9 Ruminal Fiber Breakdown and Volatile Fatty Acid (VFA) Formation in Lambs Raised with Defined Cellulolytic Microbial Populations in Sterile Isolators

Ruminal parameter	Ruminal cellulolytic population			
	Fungi	Fungi + *R. flavefaciens*	Cellulolytic bacteria	Cellulolytic bacteria + fungi
Dry matter digestion (%)				
Ryegrass	54.5 ± 3.1	62.0 ± 2.3	ND	ND
Wheat straw (48 h in sacco)	26.5 ± 0.7	32.2 ± 1.8	32.0 ± 1.7	30.5 ± 0.8
Wheat straw (72 h in sacco)	ND	ND	40.0 ± 2.1	37.8 ± 1.3
Cellulose digestion (%)[a]				
Wheat straw (48 h in sacco)	23.1	27.3	ND	ND
VFA concentration (mM)				
Total VFA	75.0 ± 4.2	93.5 ± 5.5	98.0 ± 7.2	90.5 ± 6.5
Acetate	57.0 ± 3.7	68.3 ± 5.8	70.6 ± 7.0	63.0 ± 6.1
Propionate	12.6 ± 1.1	17.8 ± 1.7	22.5 ± 3.4	21.9 ± 4.5
Butyrate	5.3 ± 0.5	6.0 ± 0.7	5.0 ± 1.0	4.0 ± 0.6

[a]Cellulose losses were determined on the combined residues after incubation in sacco.
ND = not determined.
Source: Data from G. Fonty, A.G. Williams, K. J. Cheng, F. Bonnemoy, and P. Gouet, unpublished.

Table 10 Hydrolytic Enzyme Activities[a] in the Nonadherent and Adherent Subpopulations from the Digesta Solids and the Microbial Population of the Liquor Fraction from the Ruminal Contents of Sheep Reared with Defined Cellulolytic Microorganisms in Sterile Isolators

| | Ruminal cellulolytic population | | | | | |
| | Liquor phase population | | Nonadherent population | | Adherent population | |
Enzyme	Fungi	Fungi + R. flavefaciens	Fungi	Fungi + R. flavefaciens	Fungi	Fungi + R. flavefaciens
α-L-Arabinofuranosidase	504	246	618	629	107	121
α-D-Galactosidase	314	70	323	166	30	32
β-D-Galactosidase	333	89	375	204	61	66
β-D-Cellobiosidase	65	23	61	58	12	15
α-D-Glucosidase	143	37	144	165	13	15
β-D-Glucosidase	273	100	300	149	46	46
β-D-Xylosidase	63	48	55	94	56	125
Hemicellulase	1024	2502	1190	5042	988	12683
Xylanase	1193	2344	2164	5081	1064	10843
Carboxymethylcellulase	503	484	602	1668	1649	2631
Proteinase	94	104	141	162	442	323

[a]Enzyme-specific activities in the microbial subpopulations (defined in 6,8) from sheep inoculated with fungi or fungi + R. flavefaciens are: glycosidases, nmol p-nitrophenol released/mg protein/min; polysaccharidase, nmol reducing sugar released/mg protein/h; proteinase, μg azocasein solubilized/mg protein/h.

Source: Data from G. Fonty, A. G. Williams, K. J. Cheng, F. Bonnemoy, and P. Gouet, unpublished.

R. flavefaciens were studied in the rumen of lambs isolated 24 h after birth. These lambs were then inoculated with pure cultures of *N. frontalis* MCH3 and *P. communis* FL and in a second period with *R. flavefaciens*.

Following introduction of this cellulolytic bacterial species, there was a significant increase in the disappearances of dry matter and cellulose from ryegrass hay and wheat straw, in the rumen volatile fatty acid concentrations, and in the activity of most of the polysaccharide-degrading enzymes (Tables 9 and 10).

Fungal and cellulolytic bacterial population interactions have also been investigated in the rumen of meroxenic lambs (G. Fonty, A. G. Williams, K. J. Cheng, F. Bonnemoy, and P. Gouet, unpublished data). In this experiment, lambs were transferred into sterile isolators 24 h after birth, before inoculation with a 10^{-6} dilution of rumen contents taken from conventional sheep. A bacterial cellulolytic population was established by this procedure. The animals were inoculated subsequently with pure cultures of *N. frontalis* MCH3 and *P. communis* FL. The presence of fungi in the ruminal microbial population had little effect on dry matter disappearance and slightly decreased the volatile fatty acid concentration (Table 9).

B. Fungi–Methanogenic Bacteria Interaction

Lambs introduced into a sterile isolator 24 h after birth were inoculated and raised with *N. frontalis* and *P. communis* as the only ruminal cellulolytic

Table 11 Ruminal Fiber Breakdown and Volatile Fatty Acid (VFA) Formation in Lambs Raised Gnotobiotically with Fungi or Fungi and *Mb. ruminantium* as the Cellulolytic Microbial Population

	Microbial cellulolytic population	
Ruminal parameter	Fungi	Fungi + *Mb. ruminantium* DSM 1026
Dry matter digestion (%)		
Ryegrass	55.7	56.4
Wheat straw	27.8	30.3
Cellulose digestion (%)		
Wheat straw	23.1	25.4
VFA concentration (mM)		
Total VFA	63.0	60.0
Acetate	49.7	48.6
Propionate	8.5	7.4
Butyrate	5.0	4.0

Source: G. Fonty, A. G. Williams, F. Bonnemoy, K. J. Cheng, and P. Gouet, unpublished.

Table 12 Hydrolytic Enzyme Activities[a] in the Microbial Populations Associated with the Digesta Solids and Liquor Fraction of Ruminal Contents of Gnotobiotically Raised Sheep with Ruminal Fungi or Ruminal Fungi and *Mb. ruminantium* DSM 1026 as the Cellulolytic Microbial Population

| | Liquor phase population | | Microbial population | | | |
| | | | Nonadherent population | | Adherent population | |
Enzyme	Fungi	Fungi + *Mb. ruminantium*	Fungi	Fungi + *Mb. ruminantium*	Fungi	Fungi + *Mb. ruminantium*
α-L-Arabinofuranosidase	187	138	473	417	144	103
β-D-Xylosidase	25	39	105	50	93	138
α-D-Galactosidase	43	56	128	84	41	54
β-D-Galactosidase	89	97	243	112	97	73
β-D-Cellobiosidase	10	23	46	29	15	7
α-D-Glucosidase	30	45	103	109	28	41
β-D-Glucosidase	52	125	169	152	97	123
Hemicellulase	900	3821	1544	1543	3387	12436
Xylanase	1388	4912	2003	2929	2090	13437
Carboxymethylcellulase	168	1926	3368	891	1576	6687
Proteinase	356	164	220	150	377	276

[a]For units see footnote to Table 10.
Source: G. Fonty, A. G. Williams, K. J. Cheng, F. Bonnemoy, and P. Gouet, unpublished.

microorganisms and *Mb. ruminantium* as the sole methanogenic species. The introduction of the methanogen resulted in a small increase in the extent of cellulose and dry matter digestion, although volatile fatty acid concentrations were not affected (Table 11) (92). The specific activities of the polysaccharidase enzymes in the microbial populations from the liquid phase and solid fraction of the ruminal contents (6) increased significantly after establishment of the methanogen (Table 12). However, glycohydrolase activities did not show a similar response, although in vitro the association of the fungi and a methanogen has pronounced effects on the level of enzyme activity (20). The methanogenic population did not establish effectively, and an alternative route of hydrogen transfer between the fungi and non-methanogenic hydrogenotrophic bacteria may have occurred. Such hydrogenotrophs readily establish in the developing rumen (93). It appears from the above studies that inoculation of very young animals with defined microbial populations offers considerable potential for evaluating the extent and consequences of fungal interactions with the diverse microbial groups that are normally resident within the rumen ecosystem.

ACKNOWLEDGMENTS

A. G. Williams and K. N. Joblin acknowledge the financial support of the Organization for Economic Cooperation and Development (OECD) for some aspects of the studies described. The provision of preprints of unpublished papers by colleagues is also acknowledged.

REFERENCES

1. Prins RA, van den Vorstenbosch CJ. Interrelationships between rumen microorganisms. Misc Pap Landbouwhogeschool Wageningen 1975; 11:15–24.
2. Wolin MJ. Hydrogen transfer in microbial communities. In: Bull AT, Slater JH, eds. Microbial Interactions and Communities. Vol. 1. London: Academic Press, 1982:323–356.
3. Wolin MJ, Miller TL. Microbe-microbe interactions. In: Hobson PN, ed. The Rumen Microbial Ecosystem. London: Elsevier Applied Science, 1988:343–359.
4. Forsberg CW, Lam K. Use of adenosine 5'-triphosphate as an indicator of microbiota biomass in rumen contents. Appl Environ Microbiol 1977; 33:528–537.
5. Craig WM, Broderick GA, Ricker DB. Quantitation of microorganisms associated with rumen particles. J Anim Sci (Suppl) 1983; 57:425 (Abstract).
6. Williams AG, Strachan NH. The distribution of polysaccharide-degrading enzymes in the bovine rumen digesta ecosystem. Curr Microbiol 1984; 10:215–220.
7. Czerkawski JW, Cheng K-J. Compartmentation in the rumen. In: Hobson PN, ed. The Rumen Microbial Ecosystem. London: Elsevier Applied Science, 1988: 361–385.

8. Williams AG, Withers SE, Strachan NH. Postprandial variations in the activity of polysaccharide-degrading enzymes in microbial populations from the digesta solids and liquor fractions of rumen contents. J Appl Bacteriol 1989; 66:15-26.

9. Orpin CG, Joblin KN. The rumen anaerobic fungi. In: Hobson PN, ed. The Rumen Microbial Ecosystem. London: Elsevier Applied Science, 1988:129-150.

10. Bauchop T, Mountfort DO. Cellulose fermentation by a rumen anaerobic fungus in both the absence and the presence of rumen methanogens. Appl Environ Microbiol 1981; 42:1103-1110.

11. Mountfort DO, Asher RA, Bauchop T. Fermentation of cellulose to methane and carbon dioxide by a rumen anaerobic fungus in a triculture with *Methanobrevibacter* sp. strain RA1 and *Methanosarcina barkeri*. Appl Environ Microbiol 1982; 44:128-134.

12. Fonty G, Gouet Ph, Sante V. Influence d'une bactérie méthanogène sur l'activité celluloytique et le métabolisme de deux espèces de champignons cellulolytiques du rumen *in vitro*. Résultats préliminaires. Reprod Nutr Develop 1988; 28:133-134.

13. Vogels GD, Hoppe WF, Stumm CK. Association of methanogenic bacteria with rumen ciliates. Appl Environ Microbiol 1980; 40:608-612.

14. Stumm CK, Gijzen HJ, Vogels GD. Association of methanogenic bacteria with ovine rumen ciliates. Br J Nutr 1982; 47:95-99.

15. Krumholz L, Forsberg CW, Veira DM. Association of methanogenic bacteria with rumen protozoa. Can J Microbiol 1983; 29:676-680.

16. Hillman K, Lloyd D, Williams AG. Interactions between the methanogen *Methanosarcina barkeri* and rumen holotrich ciliate protozoa. Lett Appl Microbiol 1988; 7:49-53.

17. Stewart CS, Gilmour J, McConville ML. Microbial interactions, manipulation and genetic engineering. In: New Developments and Future Prospects for Research into Rumen Function. Committee of European Communities [Report] EUR 10054, 1986:243-257.

18. Richardson AJ, Stewart CS, Campbell GP, et al. Influence of coculture with rumen bacteria on the lignocellulolytic activity of phycomycetous fungi from the rumen. Proceedings of the XIV International Congress of Microbiology. 1986; Abstract PG2-24.

19. Stewart CS, Richardson AJ. Enhanced resistance of anaerobic rumen fungi to the ionophores monensin and lasalocid in the presence of methanogenic bacteria. J Appl Bacteriol 1989; 66:85-93.

20. Joblin KN, Naylor G, Williams AG. The effect of *Methanobrevibacter smithii* on the xylanolytic activity of anaerobic rumen fungi. Appl Environ Microbiol 1990; 56:2287-2295.

21. Bernalier A, Fonty G, Gouet P. Cellulose degradation by two rumen anaerobic fungi in monoculture or in coculture with rumen bacteria. Anim Feed Sci Technol 1991; 32:131-136.

22. Joblin KN, Campbell GP, Richardson AJ, Stewart CS. Fermentation of barley straw by anaerobic rumen bacteria and fungi in axenic culture and in coculture with methanogens. Lett Appl Microbiol 1989; 9:195-197.

23. Phillips MW, Gordon GLR. Sugar and polysaccharide fermentation by rumen anaerobic fungi from Australia, Britain and New Zealand. BioSystems 1988; 21:377-383.

24. Borneman WS, Akin DE, Ljungdahl LG. Fermentation products and plant cell wall-degrading enzymes produced by monocentric and polycentric anaerobic ruminal fungi. Appl Environ Microbiol 1989; 55:1066-1073.

25. Yarlett N, Orpin CG, Munn EA, et al. Hydrogenosomes in the rumen fungus *Neocallimastix patriciarum*. Biochem J 1986; 236:729-739.

26. Wolin MJ, Miller TL. Interaction of microbial populations in cellulose fermentation. Fed Proc 1983; 42:109-114.

27. Carroll EJ, Hungate RE. Formate dissimilation and methane production in bovine rumen contents. Arch Biochem Biophys 1957; 56:525-536.

28. Lovley DR, Greening RC, Ferry JG. Rapidly growing rumen methanogenic organism that synthesizes coenzyme M and has a high affinity for formate. Appl Environ Microbiol 1984; 48:81-87.

29. Boone DR, Johnson RL, Liu Y. Diffusion of the interspecies electron carriers hydrogen and formate in methanogenic ecosystems and its implications in the measurement of Km for H_2 or formate uptake. Appl Environ Microbiol 1989; 55:1735-1741.

30. Marvin-Sikkema F, Richardson AJ, Stewart CS, et al. Influence of hydrogen-consuming bacteria on cellulose degradation by anaerobic fungi. Appl Environ Microbiol 1990; 56:3793-3797.

31. Teunissen MJ, Kets EPW, Op den Camp JHM, et al. Effect of coculture of anaerobic fungi isolated from ruminants and non-ruminants with methanogenic bacteria on cellulolytic and xylanolytic enzyme activities. Arch Microbiol 1992; 157:176-182.

32. Latham MJ, Wolin MJ. Fermentation of cellulose by *Ruminococcus flavefaciens* in the presence and absence of *Methanobacterium ruminantium*. Appl Environ Microbiol 1977; 34:297-301.

33. Williams AG, Withers SE, Joblin KN. The effect of cocultivation with hydrogen-consuming bacteria on xylanolysis by *Ruminococcus flavefaciens*. Curr Microbiol 1994; (in press).

34. Joblin KN, Williams AG. Effect of cocultivation of ruminal chytrid fungi with *Methanobrevibacter smithii* on lucerne stem degradation and extracellular fungal enzyme activities. Lett Appl Microbiol 1991; 12:121-124.

35. Williams AG. Rumen holotrich ciliate protozoa. Microbiol Rev 1986; 50:25-49.

36. Fay JP, Ovejero FMA. Effect of lactate on the *in vitro* digestion of *Agropyron elongatum* by rumen microorganisms. Anim Feed Sci Technol 1986; 16:161-167.

37. Theodorou MK, Lowe SE, Trinci APJ. The fermentative characteristics of anaerobic rumen fungi. BioSystems 1988; 21:371-376.

38. Mountfort DO, Asher RA. Production and regulation of cellulase by two strains of the rumen anaerobic fungus *Neocallimastix frontalis*. Appl Environ Microbiol 1985; 49:1314-1322.

39. Williams AG, Orpin CG. Polysaccharide-degrading enzymes formed by three species of anaerobic rumen fungi grown on a range of carbohydrate substrates. Can J Microbiol 1987; 33:418-426.

40. Williams AG, Orpin CG. Glycoside hydrolase enzymes present in the zoospore and vegetative growth stages of the rumen fungi *Neocallimastix patriciarum, Piromonas communis,* and an unidentified isolate, grown on a range of carbohydrates. Can J Microbiol 1987; 33:427–434.

41. Hébraud M, Fèvre M. Characterization of glycoside and polysaccharide hydrolases secreted by the rumen anaerobic fungi *Neocallimastix frontalis, Sphaeromonas communis* and *Piromonas communis.* J Gen Microbiol 1988; 134:1123–1129.

42. Mountfort DO, Asher RA. Production of xylanase by the ruminal anaerobic fungus *Neocallimastix frontalis.* Appl Environ Microbiol 1989; 55:1016–1022.

43. Wood TM, Wilson CA, McCrae SI, Joblin KN. A highly active extracellular cellulase from the anaerobic rumen fungus *Neocallimastix frontalis.* FEMS Microbiol Lett 1986; 34:37–40.

44. Williams AG. Hemicellulose utilization by microorganisms in the alimentary tract of ruminant and non-ruminant animals. In: Coughlan MP, ed. Enzyme Systems for Lignocellulose Degradation. London: Elsevier Applied Science, 1989:183–219.

45. Mountfort DO, Asher RA. Production of α-amylase by the rumen anaerobic fungus *Neocallimastix frontalis.* Appl Environ Microbiol 1988; 54:2293–2299.

46. Lowe SE, Theodorou MK, Trinci APJ. Cellulases and xylanase of an anerobic rumen fungus grown on wheat straw, wheat straw holocellulose, cellulose and xylan. Appl Environ Microbiol 1987; 53:1216–1223.

47. Mountfort DO, Asher RA. Role of catabolite regulatory mechanisms in control of carbohydrate utilization by the rumen anaerobic fungus *Neocallimastix frontalis.* Appl Environ Microbiol 1983; 46:1331–1338.

48. Orpin CG, Letcher AJ. Utilization of cellulose, starch, xylan and other hemicelluloses for growth by the rumen phycomycete *Neocallimastix frontalis.* Curr Microbiol 1979; 3:121–124.

49. Stewart CS, McPherson CA, Cansunar E. The effect of lasalocid on glucose uptake, hydrogen production and the solubilization of straw by the anaerobic rumen fungus *Neocallimastix frontalis.* Lett Appl Microbiol 1987; 5:5–7.

50. Stewart CS, Duncan SH, Joblin KN. Antibiotic manipulation of the rumen microflora. The effects of avoparcin and monensin on the release of tritium from labelled cellulose by *Bacteroides succinogenes* and the rumen fungus *Neocallimastix frontalis.* In: Borriello SP, Hardie JM, eds. Recent Advances in Anaerobic Bacteriology. Dordrecht: Martinus Nijhof, 1987:108–119.

51. Bernalier A, Bogaert C, Fonty G, Jouany JP. Effect of ionophore antibiotics on anaerobic rumen fungi. In: Nolan JV, Leng RA, Demeyer DI, eds. The Role of Protozoa and Fungi in Ruminant Digestion. Armidale, Australia: Penambul Books, 1989:273–275.

52. Stewart CS, Richardson AJ, Douglas RM, Rumney CJ. Hydrogen transfer in mixed cultures of anaerobic bacteria and fungi with *Methanobrevibacter smithii.* In: Belaich JP, Bruschi M, Garcia JL, eds. Microbiology and Biochemistry of Strict Anaerobes Involved in Interspecies Hydrogen Transfer. New York: Plenum Press, 1990:121–132.

53. Hungate RE. Hydrogen as an intermediate in the rumen fermentation. Arch Microbiol 1967; 59:158–164.

54. Stewart CS, Bryant MP. The rumen bacteria. In: Hobson PN, ed. The Rumen Microbial Ecosystem. London: Elsevier Applied Science, 1988:21–75.

55. Prins RA. Biochemical activities of gut microorganisms. In: Clarke RTJ, Bauchop T, eds. Microbial Ecology of the Gut. London: Academic Press, 1977:76–183.

56. Lauterbach F, Körtner C, Tripier D, Unden G. Cloning and expression of the genes of two fumarate reductase subunits from *Wolinella succinogenes*. Eur J Biochem 1987; 166:447–452.

57. Henderson C. The influence of extracellular hydrogen on the metabolism of *Bacteroides ruminicola, Anaerovibrio lipolytica* and *Selenomonas ruminantium*. J Gen Microbiol 1980; 119:485–491.

58. Greening RC, Leedle JAZ. Enrichment and isolation of *Acetitomaculum ruminis*, gen. nov., sp. nov.: acetogenic bacteria from the bovine rumen. Arch Microbiol 1989; 151:399–406.

59. Bryant MP, Small N , Bouma C, Robinson I. Studies on the composition of the ruminal flora and fauna of young calves. J Dairy Sci 1958; 41:1747–1767.

60. Genthner BRS, Davis CL, Bryant MP. Features of rumen and sewage strains of *Eubacterium limosum*, a methanol and $H_2:CO_2$-utilizing species. Appl Environ Microbiol 1981; 42:12–19.

61. Richardson AJ, Stewart CS. Hydrogen transfer between *Neocallimastix frontalis* and *Selenomonas ruminantium* grown in mixed culture. In: Belaich JP, Bruschi M, Garcia J-L, eds. Microbiology and Biochemistry of Strict Anaerobes Involved in Interspecies Hydrogen Transfer. New York: Plenum Press, 1990: 463–466.

62. Miura H, Horiguchi M, Matsumoto T. Nutritional interdependence among rumen bacteria, *Bacteroides amylophilus, Megasphaera elsdenii* and *Ruminococcus albus*. Appl Environ Microbiol 1980; 40:294–300.

63. Scheifinger CC, Wolin MJ. Propionate formation from cellulose and soluble sugars by combined cultures of *Bacteroides succinogenes* and *Selenomonas ruminantium*. Appl Microbiol 1973; 26:789–795.

64. Chen M, Wolin MJ. Influence of methane production by *Methanobacterium ruminantium* on the fermentation of glucose and lactate by *Selenomonas ruminantium*. Appl Environ Microbiol 1977; 34:756–759.

65. Williams AG, Withers SE, Joblin KN. Xylanolysis by cocultures of the rumen fungus *Neocallimastix frontalis* and ruminal bacteria. Lett Appl Microbiol 1991; 12:232–235.

66. Counotte GHM, Lankhorst A, Prins RA. Role of DL-lactic acid as an intermediate in rumen metabolism of dairy cows. J Anim Sci 1983; 56:1222–1235.

67. Bernalier A, Fonty G, Gouet P. Dégradation et fermentation de la cellulose par *Neocallimastix* sp. seul ou associé à quelques espèces bacteriènnes du rumen. Reprod Nutr Develop 1988; 28(suppl 1):75–76.

68. Bernalier A, Fonty G, Bonnemoy F, Gouet P. Degradation and fermentation of cellulose by the rumen anaerobic fungi in axenic cultures or in association with cellulolytic bacteria. Curr Microbiol 1992; 25:143–148.

69. Roger V, Grenet E, Jamot J, et al. Degradation of maize stem by two rumen fungal species *Piromyces communis* and *Caecomyces communis* in pure cultures or in association with cellulolytic bacteria. Reprod Nutr Develop 1992; 32:321–329.

70. Roger V, Bernalier A, Grenet E, et al. Degradation of wheat straw and maize stem by monocentric and polycentric rumen fungi alone or in association with rumen cellulolytic bacteria. Anim Feed Sci Technol 1993; 42:69–82.
71. Stewart CS, Duncan SH, Richardson AJ, et al. The inhibition of fungal cellulolysis by cell-free preparations from ruminococci. FEMS Microbiol Lett 1992; 97:83–88.
72. Williams AG, Withers SE, Coleman GS. Glycoside hydrolases of rumen bacteria and protozoa. Curr Microbiol 1984; 10:287–294.
73. Wallace RJ, Joblin KN. Proteolytic activity of a rumen anaerobic fungus. FEMS Microbiol Lett 1985; 29:19–25.
74. Wallace RJ, Munro CA. Influence of the rumen anaerobic fungus *Neocallimastix frontalis* on the proteolytic activity of a defined mixture of rumen bacteria growing on a solid substrate. Lett Appl Microbiol 1986; 3:23–26.
75. Williams AG, Coleman GS. The rumen protozoa. In: Hobson PN, ed. The Rumen Microbial Ecosystem. London: Elsevier Applied Science, 1988:77–128.
76. Kurihara Y, Eadie JM, Hobson PN, Mann SO. Relationship between bacteria and ciliate protozoa in the sheep rumen. J Gen Microbiol 1968; 51:267–288.
77. Coleman GS. Protozoal–bacterial interaction in the rumen. In: Nolan JV, Leng RA, Demeyer DI, eds. The Roles of Protozoa and Fungi in Ruminant Digestion. Armidale, Australia: Penambul Books, 1989:13–27.
78. Joblin KN. Bacterial and protozoal interactions with ruminal fungi. In: Akin DE, Ljungdahl LG, Wilson JR, Harris PJ, eds. Microbial and Plant Opportunities to Improve Lignocellulose Utilization by Ruminants. New York: Elsevier, 1990:311–324.
79. Williams AG, Joblin KN, Butler RD, et al. Interactions bacteries-protistes dans le rumen. Ann Biol 1993; 32:13–30.
80. Joblin KN. Isolation, enumeration and maintenance of rumen anaerobic fungi in roll tubes. Appl Environ Microbiol 1981; 42:1119–1122.
81. Orpin CG. Studies on the defaunation of the ovine rumen using dioctyl sodium sulphosuccinate. J Appl Bacteriol 1977; 43:309–318.
82. Fonty G, Joblin KN. Rumen anaerobic fungi: their role and interactions with rumen micro-organisms in relation to fibre digestion. In: Tsuda T, Sasaki Y, Kawashima R, eds. Physiological Aspects of Digestion and Metabolism in Ruminants. New York: Academic Press, 1991:655–680.
83. Demeyer DI. Rumen microbes and digestion of plant cell walls. Agric Environ 1981; 6:295–337.
84. Orpin CG. The role of ciliate protozoa and fungi in the rumen digestion of plant cell walls. Anim Feed Sci Technol 1983/84; 10:121–143.
85. Eadie JM. Inter-relationships between certain rumen ciliate protozoa. J Gen Microbiol 1962; 29:579–588.
86. Jouany JP. Effects of diet on populations of rumen protozoa in relation to fibre digestion. In: Nolan JV, Leng RA, Demeyer DI, eds. The Role of Protozoa and Fungi in Ruminant Digestion. Armidale, Australia: Penambul Books, 1989: 59–74.
87. Romulo BH, Bird SH, Leng RA. The effects of defaunation on digestibility and rumen fungi counts in sheep fed high-fibre diets. Proc Aus Soc Anim Prod 1987; 16:327–330.

88. Romulo BH, Bird SH, Leng RA. Effects of defaunation and protein supplementation on intake, digestibility, N retention and fungal numbers in sheep fed straw-based diets. In: Nolan JV, Leng RA, Demeyer DI, eds. The Role of Protozoa and Fungi in Ruminant Digestion. Armidale, Australia: Penambul Books, 1989:285–288.
89. Williams AG, Withers SE. Effect of ciliate protozoa on the activity of polysaccharide-degrading enzymes and fibre breakdown in the rumen ecosystem. J Appl Bacteriol 1991; 70:144–155.
90. Fonty G, Roussel O, Gouet Ph, Chavarot M. Activité cellulolytique in vivo de *Bacteroides succinogenes, Ruminococcus flavefaciens* et *Ruminococcus albus* dans le rumen d'agneaux placés en isolateurs 24 heures après la naissance. Reprod Nutr Develop 1988; 28:135–136.
91. Fonty G, Gouet Ph. Establishment of microbial populations in the rumen of lambs. Utilisation of an animal model to study the role of the different cellulolytic microoganisms *in vivo*. In: Nolan JV, Leng RA, Demeyer DI, eds. The Role of Protozoa and Fungi in Ruminant Digestion. Armidale, Australia: Penambul Books, 1989:39–49.
92. Grenet E, Fonty G, Barry P. SEM study of the degradation of maize and lucerne stems in the rumen of gnotobiotic lambs harbouring only fungi as cellulolytic microorganisms. In: Nolan JV, Leng RA, Demeyer DI, eds. The Role of Protozoa and Fungi in Ruminant Digestion. Armidale, Australia: Penambul Books, 1989:265–267.
93. Dore J, Rieu-Lesme F, Fonty G, Gouet Ph. A preliminary study of the nonmethanogenic hydrogenotrophic microflora in the rumen of new-born lambs. Ann Zootech 1991; 41:82 (Abstract).

8

Effects of Diet on the Fungal Population of the Digestive Tract of Ruminants

G. FONTY and E. GRENET
Institut National de la Recherche Agronomique
Saint Genes Champanelle, France

I. INTRODUCTION

The high cellulolytic, hemicellulolytic, and proteolytic activity of rumen anaerobic fungi observed in vitro in pure cultures or cocultures, their physical association with the lignocellulosic tissues of plant particles, observed in vivo, and their ability to produce a large panel of plant cell-wall–degrading enzymes suggest that the fungi are able to play an important role in vivo (for references see Refs. 1,2). The contribution of these microorganisms in the rumen to the degradation and fermentation of feed components and to the nutrition of the host animal depends of course on their number. The quantitative and qualitative composition of the mycoflora, like that of the other rumen populations, depends on the abiotic conditions in the rumen and consequently on the diet given to the animals.

II. ESTABLISHMENT OF FUNGI IN THE RUMEN OF THE YOUNG ANIMAL

The fungi appear in the rumen rapidly after birth, before the organ becomes functional. They were observed by Fonty et al. (3) 8 to 10 days after birth in lambs flock-reared with their mothers. The microorganisms were present in all animals studied and consisted mainly of *Neocallimastix frontalis. Sphae-*

romonas communis was isolated in a few animals. The fungi do not require plant material to become established, since at this early age the rumen content is composed entirely of mucus and desquamated epithelial cells (3,4). However, the subsequent development of the fungal population is largely dependent on the feed distributed at weaning. In 80% of the lambs studied by Fonty et al. (3) the fungi had disappeared after the age of 3 weeks, once the animals had access to a concentrated solid feed. In contrast, the fungal flora was maintained in animals receiving dehydrated lucerne hay (5).

Whereas fungi do not need plant fibers to become established, they can be transmitted only if there are close contacts between the newborns, their mothers, and other ruminants. They have been observed in the saliva and feces (6), and so it is likely that transmission is made by direct oral contact, by aerosol, or by fecal contamination (7).

III. EFFECTS OF DIET ON THE RUMEN FUNGAL POPULATION

Several authors have shown that the fungal population varies according to the composition of the diet (8–11). However, in view of the absence of any reliable technique for assessing the fungal biomass in vivo (see Chapter 10) and of the limits of classical microbiological techniques, it is sometimes very difficult to make a quantitative comparison of fungal populations in animals fed different diets. Likewise, it is difficult to assess precisely the influence of feed on the quantitative composition of the fungal flora, since there exist no selective media for counting the different genera or species. Until now, three main methods have been used to make quantitative comparisons of fungal populations: counts of free zoospores in rumen liquid phase (12); of the percentage of plant particles colonized by the fungi (11); or of sporocysts on soybean hulls (10) or on agar fragments supplemented with sugar (13). The soybean hulls and the agar fragments are introduced into the rumen in nylon bags for determined periods. Despite the reservations expressed above, it is clear that certain feeds are more favorable to the development of fungi than others.

A. Diets Favorable to the Development of the Fungal Flora in the Rumen

The number of fungi in the rumen, as determined by counts of zoospores present in a milliliter of ruminal fluid, varies according to diet. A range of 0 to 4×10^4 zoospores ml^{-1} of ruminal fluid has been observed (10,11). Diets favorable or very favorable to the development of fungi, for which numbers can vary between 6×10^3 and 4×10^4 (Table 1), are either made up entirely of forage, such as grass silage, ryegrass at earing stage, or lucerne hay, or consist predominantly of forage, such as straw- or maize silage-based diets.

Table 1 Diets Favorable for the Development of Rumen Fungi

Diet	Composition	Number of zoospores ml^{-1} fluid ($\times$ 10^3)	pH rumen fluid
Grass silage[a]	8 kg late fescue silage	40	6.8
Ryegrass: ear emergence stage[a]	Grazing perennial ryegrass	40	6.8
Maize silage +monensin[b]	10 kg maize silage +400 mg monensin	29	6.2
Straw[a]	4.8 kg wheat straw +2 kg ground maize +0.07 kg urea +0.2 kg tanned oil meal	20	7.0
Lucerne hay[a]	16.2 kg lucerne hay	11	6.5
Maize silage + concentrate[a]	5.6 kg maize silage +4.6 kg energy concentrate +0.4 kg tanned rapeseed and soya bean meal (50/50)	8	6.7
Maize silage[a]	12.1 kg maize silage +0.14 kg urea	7	6.7
Lucerne hay[b]	10 kg lucerne hay	6	6.7

[a]From Ref. 10.
[b]From Ref. 11.

All these different diets provide the animal with plant cell-walls as shown by the crude fiber content and neutral detergent fiber content (Table 2), and, as noted by Bauchop (14), they supply quantities of stems containing lignified tissues. The various electron-microscopic studies made on anaerobic fungi all show that the microorganisms attach essentially to thick-walled or lignified tissues (14,15,16) and that diets supplying the latter in large quantities are favorable to their growth. The fact that the rumen fungi become attached to lignified tissues in both conventionally reared lambs and gnotobiotic animals harboring only fungi as cellulolytic microorganisms shows that there is a process of adaptation (17). Thus, in gnotobiotic animals harboring only fungi as cellulolytic microorganisms, in which there is no competition between cellulolytic bacteria and fungi for attachment sites on the plant tissues, they preferentially attach to the lignified tissues.

Bauchop (14) observed that a diet of pelleted lucerne hay was far less favorable to fungal development than one of chopped lucerne. The plant tissue content of the two diets was the same, but the pellets of lucerne hay had a shorter transit time in the rumen because of their smaller particle size. The life cycle of rumen fungi has been estimated at about 24 h (14); lignified tissues, being resistant to degradation, have an average transit time in the

Table 2 Chemical Composition of Feeds Given to the Cows in the experiments of Grenet et al.[a]

Feed	Content in % of dry matter			
	Ashes	Crude protein	Crude fiber	NDF
Maize silage				
No. 1	6.0	8.3	19.8	
No. 2	5.0	7.4	18.0	
No. 3	5.6	9.0		42.5
Lucerne hay				
No. 1	9.1	15.7	25.1	
No. 2	7.3	11.6		56.9
No. 3	9.9	18.5		40.2
Grass silage	8.0	10.4	35.1	
Straw	11.8	5.6	38.2	
Beets				
No. 1	8.2	7.6	7.4	
No. 2	5.6	7.9	6.1	
No. 3	13.8	14.5		26.1
Whey	8.0	12.8	0.3	
Beet pulps	7.7	9.5	2.7	
Barley	3.6	14.7	4.9	
Maize grain	3.4	11.8	2.3	
Ryegrass				
Leafy stage	12.5	18.1	18.2	
Ear-emergence stage	10.0	12.9	25.8	
Tanned soy bean and repeseed oil meal (50/50)	7.5	46.0	8.1	
Energy concentrate	7.5	14.0	9.2	

[a]Refs. 10 and 11. Composition of maize silage, lucerne hay, and beets was determined on three different occasions.
NDF = neutral detergent fiber.

rumen of a day unless, as for pelleted feed, their particle size is small, in which case it is shorter (18). Diets with the longest transit time are therefore the most favorable to the growth of rumen anaerobic fungi.

B. Diets Unfavorable to the Development of Fungal Flora in the Rumen

Diets weakly favorable or unfavorable to the development of the fungal population (Table 3) are those containing young forage poor in stems, such as ryegrass at grazing; those containing very little forage but rich in starch, such as barley diets, or rich in soluble sugars, such as beet diets, which contain

Table 3 Diets Unfavorable for the Development of Rumen Fungi

Diet	Composition	Number of zoospores ml^{-1} ($\times 10^3$)	pH rumen fluid
Ryegrass: leafy stage[a]	Grazing perennial ryegrass	4.0	6.6
Beets in 6 meals[a]	8.2 kg fodder beets fed in 6 equal meals per day +6.2 kg lucerne hay	1.5	6.8
Lucerne hay + whey[b]	4 kg lucerne hay + 6 kg mild whey powder	0.7	5.9
Beets in 1 meal[a]	8.2 kg fodder beets +7.4 kg lucerne hay	0.2	5.4
Barley[a]	6.8 kg ground barley + 1.7 kg maize silage + 0.4 kg tanned oil meal	0	6.5
Whey[a]	7.8 kg mild whey powder + 0.9 kg beet pulp + 6.2 kg lucerne hay	0	5.2

[a]From Ref. 10.
[b]From Ref. 11

saccharose; and those, containing large amounts of whey, which provide lactose. The number of zoospores observed with these diets ranged from 0 to 4×10^3 ml^{-1} of ruminal fluid. The pH of the rumen tends to be lower with these diets than with those that are favorable, between 5.2 and 6.8. The arrival of large quantities of rapidly fermentable sugar in the rumen decreases the pH. Orpin (19) observed in vitro that the production of zoospores by *Piromonas* was considerably less at pH 5.5, so that a lower pH might be a causative factor in the decrease in the fungal population. Moreover, the presence of large amounts of soluble sugar in the ruminal fluid, which may partly inhibit the adhesion of cellulolytic bacteria to plant particles (20), may also prevent the germination of fungal zoospores on plant tissues by saturating the spores' adhesion sites (21). This last factor plays a considerable part in the decrease in the fungal population, along with the presence of small amounts of lignified tissues, whose importance to the adhesion of fungi to plant particles is well documented. Addition of rapeseed oil to the diet (1 kg/d) leads also to a considerable decrease in the fungal population (G. Fonty et al., unpublished data). The mechanism of this inhibition has not been elucidated. Like protozoa (22), fungi may be sensitive to the polyunsaturated fatty acids.

C. Ability of Anaerobic Fungi to Colonize Plant Particles

Grenet et al. (10) studied the colonization of plant fragments placed in nylon bags that were introduced into the rumen of fistulated animals. Colonization depended on the nature of the fragments and on the diet of the animal. Lucerne and maize stems and soybean teguments introduced into the rumen in nylon bags were abundantly colonized when the number of zoospores in the rumen was high, i.e., with diets favorable to the development of the fungal population (Table 4). The fungi became attached to the xylem of the lucerne stem, to the perivascular fibers of the maize stem, or to the palisade layer of the soybean tegument. Bauchop (14) also observed an abundant colonization of the stems and leaves of lucerne and graminae. In contrast, with an unfavorable diet such as ryegrass at grazing, the three substrates were little colonized, and the sporangia on the soybean teguments had reached a size of only 30 μm after 24 h, compared with 100 μm achieved with other diets (Table 5). Thus, for a given substrate, lucerne hay, grass silage, and straw favored the growth of sporangia whose size after 24 h was greater than with other diets. The exception was beet pulp, which is rarely colonized by anaerobic fungi. Sporangia have been observed in small numbers attached to

Table 4 Colonization of Substrates in Sacco and Mean Size (Length and Width in μm) of About 10 Sporocysts After a 24-h Stay in the Rumen of Cows Fed Diets Favorable for the Development of Fungi

Diet	Colonization of substrates[a]/size of sporocysts			
	Beet pulp	Lucerne stems	Maize stalks	Soybean teguments
Grass silage	+	+ + +	+	+ +
	30 × 26	92 × 75	32 × 25	98 × 70
Ryegrass:ear	0	+ + +	+	+ +
emergence stage		65 × 32	30 × 20	72 × 44
Straw	+	+ + +	+ +	+ + +
	46 × 43	86 × 59	57 × 54	124 × 64
Maize silage +	0	+ +	+ +	+ + +
concentrate		62 × 31	44 × 20	113 × 55
Maize silage	+	+ +	+ +	+ + +
	34 × 21	57 × 34	42 × 23	80 × 44
Lucerne hay	0	+ + +	+	+ +
		93 × 61	77 × 38	120 × 59

[a]0, no sporocyst; +, a few sporocysts; + +, numerous sporocysts; + + +, substrates covered with sporocysts.
Source: From Ref. 10.

Table 5 Colonization of Substrates in Sacco and Mean Size (Length and Width in μm) of About 10 Sporocysts in the Rumen of Cows Fed Diets Unfavorable for the Development of Fungi

Diet	Residence time (h)	Colonization of substrates[a]/size of sporocysts			
		Beet pulp	Lucerne stems	Maize stalks	Soybean teguments
Ryegrass: leafy stage	24	0	+[b]	+[b]	+ 29 × 30
Beets in 6 meals	24	0	+ + 59 × 31	+ 31 × 17	+ 59 × 28
Beets in 1 meal	24	0	0	0	0
Barley	24	0	+ 108 × 61	+ 66 × 32	0
Whey	24	0	0	0	0

[a]0, no sporocyst; +, a few sporocysts; + +, numerous sporocysts; + + +, substrates covered with sporocysts.
[b]These sporocysts were not measured.
Source: From Ref. 10.

the xylem with three diets only: maize silage, grass silage, and straw (Table 4). The chemical composition of beet pulp, which is rich in cellulose, hemicellulose, and pectins (16), is not favorable to the adhesion and growth of fungi. It would seem that these microorganisms are not capable of degrading pectins, whose presence in the beet pulp cell-walls is unfavorable to their development. After 24 h there are differences in the size of the sporangia on the different substrates, according to the composition of the diet. Sporangia are bigger on soybean teguments, a little smaller on lucerne stems, and smaller still on maize stems. The composition of the substrate therefore affects the size of the sporangia. The composition of soybean teguments, which are rich in cellulose and hemicellulose (16), is favorable to the growth of rumen fungi. The crude protein content of lucerne and soybean teguments is higher than that of maize (23), which may be a favorable factor in the development of fungi, but this does not rule out the possibility that other growth factors are involved.

IV. EFFECT OF DIET ON THE FUNGAL POPULATION IN THE DUODENUM, CECUM, AND FECES

Although anaerobic fungi have been found in sheep saliva and in the digestive tract and feces of numerous herbivores (24), very few studies, either quantitative or qualitative, have been made on the fungal flora of the in-

Table 6   Number of Zoospores in the Duodenum, Cecum, and
Feces of a Cow Fed Different Diets

Diet	No. of zoospores $(\times 10^{-2}\,g^{-1}$ of digestive contents)		
	Duodenum	Cecum	Feces
Maize silage + monensin	3.8	19.3	51.0
Lucerne hay	2.5	5.1	5.5
Lucerne hay + monensin	3.3	7.7	2.9
Lucerne hay + beets	2.8	12.0	39.0
Lucerne hay + whey	0.1	1.0	5.1

Source: From Ref. 11.

testine. However, it would seem that in ruminants the influence of feed content is not confined to the rumen but is felt elsewhere (11).

In the cow, the diets that were the most favorable to the development of fungi in the rumen were also those that most actively promoted growth in the duodenum, cecum, and feces. In these three biotopes, however, the number of fungi as determined by zoospore counts was markedly lower than in the rumen, especially in the duodenum (Table 6). On average, numbers were two to three times lower in the duodenum than in the cecum. Highest counts were observed with maize silage plus monensin and with lucerne hay plus beets.

Table 7   Proportion of Particles Colonized by Anaerobic Fungi
in the Intestine and Feces of Cows

Diet	% of particles colonized by anaerobic fungi		
	Duodenum	Cecum	Feces
Maize silage + monensin	29.1	26.3	35.0
Lucerne hay + whey[a]	17.7	22.1	16.7
Lucerne hay + whey[b]	27.6	48.1	27.1
Lucerne hay[c]	21.0	24.9	18.9
Lucerne hay[d]	32.5	18.8	19.2
Lucerne hay[e]	34.2	37.3	35.4
Lucerne hay + monensin	31.5	41.6	35.9
Lucerne hay + beets	19.5	11.2	4.8

[a,b]Rumen samples taken week 3 and 4, respectively, after the change of diet.
[c,d,e]Rumen samples taken day 4, 8, and 18, respectively, after the change
of diet.
Source: From Ref. 11.

For a given diet the proportion of plant particles colonized by the fungi was similar in the three parts of the digestive tract (Table 7). In a given digestive compartment no significant difference was observed between the diets. With the exception of the beet diet, the percentage of particles colonized ranged between 20% and 40%. Qualitative differences have been observed between the fungal flora of the intestine and that of the rumen (11). The rumen population was essentially composed of species possessing filamentous rhizoids, *Neocallimastix* spp. and *Piromonas* spp. The intestinal population was more diversified, especially in the cecum and feces (Table 8). Colonies of nonrhizoidal species resembling *Sphaeromonas* spp. were found in numbers equal to those of the rhizoidal species. With a diet of maize silage plus monensin, the population was made up solely of nonrhizoidal species.

In sheep fed a forage diet (25), fungi were absent in the duodenum but were present in the feces at a level between 3.2×10^3 and 16×10^3 zoospores/g. *Caecomyces* (*Sphaeromonas*) constituted 75% of the fecal population, and the rhizoïdal monocentric species only 25%. *Piromyces* (*Piromonas*) *mae* was the dominant rhizoïdal population. No polycentric species were found. The rumen fungal population of these sheep was composed of 70% rhizoïdal monocentric species (*Neocallimastix, Piromyces*), 27% nonrhizoïdal species (*Caecomyces*), and 3% polycentrics (*Orpinomyces*).

V. CONCLUSION

Despite the imperfections in the techniques for estimating the number of fungi, it is clear that the fungal population varies considerably in size according to the composition of the feed ingested by the host animal. The small numbers of fungi, or even their absence, observed with easily digestible diets show that fungi are not essential per se for the digestion of ruminants. In contrast, it is likely that they play an important role in the degradation of

Table 8 Dominant Morphological Type in Colonies Developed in Roll Tubes from Samples of Rumen, Duodenal, and Cecal Contents, and Feces

Diet	Rumen	Duodenum	Cecum	Feces
Maize silage + monensin	R	NR	NR	NR
Lucerne hay + whey (week 4)	R	R	r(60), nr(40)	r(75), nr(25)
Lucerne hay (day 8)	R	R	r(80), nr(20)	r(80), nr(20)
Lucerne hay + monensin	R	R	r(50), nr(50)	r(50), nr(50)
Lucerne hay + beets	R	r(60), nr(40)	r(70), nr(30)	r(60), nr(40)

R = 100% of colonies of the rhizoidal type; NR = 100% of colonies of the nonrhizoidal type; r(80) = 80% of colonies of the rhizoidal type; nr(20) = 20% of colonies of the nonrhizoidal type.
Source: From Ref. 11.

plant cell-walls, since they are found in large numbers in animals fed rations rich in forages. The fact that they possess cellulolytic, hemicellulolytic, and proteolytic activity as well (1,2) should give them a clear ecological advantage over cellulolytic bacteria in fiber-rich diets. The presence of fungi in the rumen of the preruminant that has not yet begun to ingest solid feed raises the question of whether these microorganisms have an ecological niche other than that of the degradation of the polymers of plant cell walls.

REFERENCES

1. Orpin CG, Joblin KN. The rumen anaerobic fungi. In: Hobson PN, ed. The Rumen Microbial Ecosystem. London: Elsevier Applied Science, 1988:129–150.
2. Fonty G, Joblin KN. Rumen anaerobic fungi: their role and interactions with other rumen microorganisms in relation with fibre digestion. In: Tsuda T, Sasaki Y, Kawashima R, eds. Physiological Aspects of Digestion and metabolism in Ruminants. San Diego: Academic Press, 1990:655–680.
3. Fonty G, Gouet Ph, Jouany JP, Senaud J. Establishment of the microflora and anaerobic fungi in the rumen of lambs. J Gen Microbiol 1987; 133:1835–1843.
4. Jayne-Williams DJ. The bacterial flora of the rumen of healthy and bloating calves. J Appl Bacteriol 1979; 57:271–283.
5. Fonty G, Gouet Ph. Establishment of microbial populations in the rumen. Utilization of an animal model to study the role of the different cellulolytic microorganisms in vivo. In: Nolan JV, Leng RA, Demeyer DI, eds. The Roles of Protozoa and Fungi in Ruminant Digestion. Armidale, Australia: Penambul Books, 1989:39–49.
6. Lowe SE, Theodorou MK, Trinci APJ. Isolation of anaerobic fungi from saliva and faeces of sheep. J Gen Microbiol 1987; 133:1824–1834.
7. Orpin CG. Ecology of rumen anaerobic fungi in relation to the nutrition of the host. In: Nolan JV, Leng RA, Demeyer DI, eds. The Roles of Protozoa and Fungi in Ruminant Digestion. Armidale, Australia: Penambul Books, 1989:29–38.
8. Bauchop T. The anaerobic fungi in rumen fibre digestion. Agric Environ 1981; 6:339–348.
9. Bauchop T. Colonisation of plant fragments by protozoa and fungi. In: Nolan JV, Leng RA, Demeyer DI, eds. The Roles of Protozoa and Fungi in Ruminant Digestion. Armidale, Australia: Penambul Books, 1989:83–95.
10. Grenet E, Breton A, Barry P, Fonty G. Rumen anaerobic fungi and plant substrates colonization as affected by diet composition. Anim Feed Sci Technol 1989; 26:55–70.
11. Grenet E, Fonty G, Jamot J, Bonnemoy F. Influence of diet and monensin on development of anaerobic fungi in the rumen, duodenum, caecum and feces of cows. Appl Environ Microbiol 1989; 55:2360–2364.
12. Joblin KN. Isolation, enumeration and maintenance of rumen anaerobic fungi in roll tubes. Appl Environ Microbiol 1981; 42:1119–1122.
13. Ushida K, Tanaka H, Kojima Y. A simple in situ method for estimating fungal population size in the rumen. Lett Appl Microbiol 1989; 9:109–111.

14. Bauchop T. The rumen anaerobic fungi: colonizers of plant fibre. Ann Rech Vet 1979; 38:148–158.
15. Akin DE, Rigsby LL. Mixed fungal populations and lignocellulosic tissues degradation in the bovine rumen. Appl Environ Microbiol 1987; 53:1987–1995.
16. Grenet E, Barry P. Etude microscopique de la digestion des parois végétales des téguments de soja et de colza dans le rumen. Repr Nutr Develop 1987; 27:246–248.
17. Grenet E, Fonty G, Barry P. SEM study of the degradation of maize and lucerne stems in the rumen of gnotobiotic lambs harbouring only fungi as cellulolytic microorganisms. In: Nolan JV, Leng RA, Demeyer DI, eds. The Roles of Protozoa and Fungi in Ruminant Digestion. Armidale, Australia: Penambul Books, 1989:265–267.
18. Demarquilly C, Journet M. Valeur alimentaire des foins condensés. Influence de la nature du foin et de la finesse de broyage sur la digestibilité et la quantité ingérée. Ann Zootech 1967; 16:123–150.
19. Orpin CG. The rumen flagellate *Piromonas communis*: its life-history and invasion of plant material in rumen. J Gen Microbiol 1977a; 99:107–117.
20. Roger V. Fonty G, Komisarczuk S, Gouet Ph. Effects of physicochemical factors on the adhesion to cellulose Avicel of the ruminal bacteria *Ruminococcus flavefaciens* and *Fibrobacter succinogenes*. Subsp. *succinogenes*. Appl Environ Microbiol 1990; 56:3081–3087.
21. Orpin CG, Bountiff L. Zoospore chemotaxis in the rumen phycomycete *Neocallimastix frontalis*. J Gen Microbiol 1978; 104:113–122.
22. Van Nevel CS, Demeyer DI. Manipulation of Rumen Microbes. In: Hobson PN, ed. The Rumen Microbial Ecosystem. London: Elsevier Applied Science, 1988:387–444.
23. Institut National de la Recherche Agronomique. Amilentation des Bovins, Ovins et Caprins. Versailles, France: INRA Publications, 1988.
24. Milne A, Theodorou MK, Jordan MG, et al. Survival of anaerobic fungi in feces, in saliva and in pure culture. Exp Mycol 1989; 13:27–37.
25. Breton A, Confesson I, Dusser M, Gaillard-Martinie B. Comparaison du peuplement fongique du rumen duodénum et des fèces de mouton. Ann Zootech 1994. In press.

9

The Nucleic Acids
of Anaerobic Fungi

ALAN G. BROWNLEE

Division of Animal Production
CSIRO
Prospect, New South Wales, Australia

I. INTRODUCTION

Microbial nucleic acids have been a popular target for investigators because they are accessible macromolecules with a high information content and are present in all cells. The information contained in the sequence of nucleic acids is the source of the distinctive attributes that characterize any organism. This chapter will discuss some of the features of the nucleic acids, primarily deoxyribonucleic acid (DNA), of anaerobic gut fungi. What is known of the ribonucleic acids (RNA) is derived largely from analysis of their encoding genes. Studies in this area have revealed some surprising features of these organisms as well as information crucial to their proper taxonomic classification and an understanding of their evolutionary history and biology.

II. ISOLATION OF ANAEROBIC-FUNGAL DNA

There are many published methods for the purification of fungal DNA (1), but few are suited to efficient extraction of DNA from anaerobic fungi (2). Complications are introduced by the very low DNA content as a proportion of weight: my estimates of the DNA content of 3- to 6-day old cultures of *Neocallimastix* are 2.7–3.2 μg/mg dry weight or 0.32% by weight determined by a spectrofluorometric method (3) on sonicated extracts. This value is probably an overestimate, because of the preferential interaction of the fluorescent dye with adenine and thymine bases in DNA (Section III). The DNA content of other types of anaerobic fungi seems equally low (A. G. Brownlee, unpublished results). DNA extraction is also hampered by the presence of a considerable quantity of a glycogenlike storage polysaccharide that persists through most stages of DNA purification (2,4). The unusual base composition of the DNA (Section III) must be considered if purification includes conventional cesium chloride density gradient centrifugation, if phenol is employed during purification (5), or if steps are not included to rapidly inactivate endogenous nucleases. The small-scale purification method (2) appears to work well with all strains examined in the author's laboratory and in other laboratories where problems had been encountered previously (J. Doré, personal communication). The use of phenol is intentionally avoided, and the method can readily be scaled up. As with some other DNA purification methods, the most convenient starting material is freeze-dried mycelium ground to a fine powder with a mortar and pestle. Alumina powder can be used to assist the grinding. The DNA within this lyophilized mass (ground or intact) seems quite stable during lengthy storage at −20°C and is in a convenient form for interlaboratory exchange. DNA extraction from zoospores of anaerobic fungi has been described (6). This method combined an acetone pretreatment of the spores with cell-wall removal using lytic enzymes. The DNA yield was reported to be 61% ± 7%, and the DNA content of zoospores was estimated at 2.2 ± 0.6 pg (6). Small quantities of spores could be processed.

III. DNA BASE COMPOSITION OF ANAEROBIC FUNGI

One of the properties that characterize DNA, and hence the organism, is its base composition, more particularly its guanine-plus-cytosine (G + C) content (expressed as mol%). With the exception of the DNA encoding ribosomal RNA (rRNA) (see below), fungi in general do not carry much noncoding or repetitive DNA (7), and the genomic G + C content is subject to some variation. Kurtzman (8) has discussed how some groups of fungi can be broadly classified and differentiated according to their %G + C, but the taxonomic value of these measures is largely confined to excluding the identity of strains because of the limited range of values normally encountered (approximately 30–70 mol%). The genomic G + C content of the anaerobic fungi has turned out to be unique in this respect; as a group they are clearly different from any other type of fungus. An isolate of the monocentric *Neocallimastix* (strain LM-2) has been examined in some detail, and its DNA has a G + C value of 18% (9). This value includes a contribution by the highly repetitive genes encoding the rRNA, and when allowance is made for this fraction, the unique sequence DNA is perhaps as low as 12–13% G + C (9). There is now some confirmation of this extraordinary composition in other strains of anaerobic fungi (10,11,12) (see Table 1). Evidence was presented that the adenine and thymine (A + T)-rich sequences are widely dispersed in the genome of strain LM-2 (9), and on the basis of the high frequency of sites for those restriction endonucleases with A + T-rich recognition sequences, the same appears to be the case for the DNA of the other strains of multiflagellate and uniflagellate anaerobic fungi examined (A. G. Brownlee, unpublished results). These results indicate that the anaerobic fungi have the most A + T-rich genomes of any organism identified so far — to such an extent that they test the limits of the physicochemical methods commonly used in their analysis. Other chytrid fungi are more orthodox in DNA base composition (Table 1). Since the taxonomic relatedness of the anaerobic fungi to the Chytridiomycota is now strongly supported on molecular as well as morphological grounds (Section VII), it would seem that some aspect of the unique enviroment of these fungi is responsible for their extreme base composition (see discussion in Ref. 9).

Base composition analysis following separation of the constituent bases by high-performance liquid chromatography (9) did not reveal any detectable

Table 1   DNA Base Composition of Anaerobic and Related Fungi

Organism	mol% G + C	Ref.
Order Blastocladiales		
Blastocladiella emersonii	66	13
Allomyces arbuscula	62.2	8
Blastocladia ramosa[a]	52.4	
Blastocladia pringsheimii[a]	56.9	
Order Chytridiales		
Rhizophydium sp.	50.5	13
Order Spizellomycetales		
Rhizophlyctis rosea	44	13
Neocallimastix LM-2	18	9
N. frontalis[b]	17.4, 16.2	10
N. joyonii	14.6	10
Caecomyces communis[b]	22.1, 19.9	10
Anaeromyces mucronatus[b]	15.7, 14.9	10
Piromyces communis	15.5	10

[a]G + C content was determined in the authors's laboratory using high-performance liquid chromatography (9) (A. G. Brownlee, unpublished results).
[b]Two determinations were reported for DNA prepared by two methods.
G + C = guanosine plus cytosine.

amounts of modified bases, in particular 5-methylcytosine. Although 5-methyl-cytosine is not commonly found in fungal DNA (14), its presence has been documented (15,16).

IV. CONSEQUENCES OF THE EXTREME BASE COMPOSITION

By analogy with the effects of base composition on the protein encoding se-quences in the DNA of other organisms (17,18), the highly A + T-rich genome of the anaerobic fungi might be expected to affect the amino acid composi-tion of the encoded proteins. This does, in fact, appear to be the case (19). The codon bias resulting from the low G + C content leads to a depletion in the total cellular content, within protein, of those amino acids coded for by G + C-rich codons (e.g., arginine, proline, glycine) and an overabundance of those amino acids coded for by A + T-rich codons (e.g., lysine, phenyla-lanine, tyrosine). The correlation between these effects is sound enough (19) for a range of bacteria and eucaryotic microbes, including fungi, that the amino acid composition of a number of strains of anaerobic fungi can ap-parently be used to predict the presence of an underlying A + T-rich genome. Whatever the basis for the low G + C content, there are potential nutritional consequences for the animals harboring these organisms, particularly for ruminants, in which the rumen microflora make a major contribution to the amino acid supply.

There might be advantages for molecular studies involving the anaerobic fungi resulting from their A + T-rich genome. The design of mixed oligo-nucleotide probes, based on sequence information from other sources, can be simplified where assumptions can be made about restricted codon usage (20). This is likely to assist in the isolation of genes from these fungi and, presumably, in the identification of open reading frames in regions of se-quenced genomic DNA. The predictable limits placed on the frequencies of recognition sites for restriction endonucleases in the genome (9) can be put to use for the cloning of large or small fragments of DNA. Certain restric-tion enzymes that recognize A + T-containing sequences can be used to cut the genome frequently, into quite small, "gene-size" pieces (9,21; and refer-ences therein).

V. IS MITOCHONDRIAL DNA PRESENT?

Mitochondria, as recognizable organelles, are absent from the anaerobic fun-gi. The evolutionary origin of these fungi (see below) suggests that their an-cestors possessed mitochondria and that these were lost from the line leading to the anaerobes. How this was accomplished biologically is a mystery, but there remains the formal possibility that a vestigial organelle has remained undetected, in which case the accompanying DNA should still be present

(since many mitochondrial functions are encoded by an autonomously re-
plicating genome). Another possibility is that some mitochondrially encoded
functions have been transferred to the nuclear genome (22). DNA-hybridiza-
tion experiments using yeast mitochondrial DNA as a probe (A. G. Brownlee,
unpublished results) have failed to detect any homologous sequences in total
DNA of *Neocallimastix*. A far more sensitive technique, the polymerase chain
reaction (Section VI.C), using conserved primers directed toward the ribosomal large subunit RNA gene of the fungal mitochondrial genome, failed
to reveal any amplification product within *Neocallimastix* DNA (T. White,
personal communication).

Mitochondrial DNA has served an important function for biologists studying the evolution and genetic diversity of fungi (22). The likely absence of this
genome from anaerobic fungi removes a potentially useful macromolecule
from consideration in molecular studies of these organisms, and increased
emphasis will have to be placed on those common features that they have
retained, such as the rRNA genes (Section VI.C).

VI. IDENTIFICATION OF SPECIFIC DNA SEQUENCES

A. Cloned Anonymous Sequences

An analysis of randomly selected clones of *Neocallimastix* genomic DNA in
the bacteriophage cloning vector M13 revealed a relatively complex organization of A + T-rich sequences, including the presence of short direct and
inverted repeats and regions of dyad symmetry (9). A more extensive study
suggests that this type of organization may be representative of several strains
of this genus (23). In this report, an anonymous repetitive sequence from
Neocallimastix strain LM-2 was cloned in a plasmid vector and used as a
DNA probe to screen isolates of anaerobic fungi. The results of DNA–DNA
hybridization indicated that the isolated repeat sequence (designated pLM-6)
was common in the genome of all the *Neocallimastix* strains examined, producing a characteristic pattern of hybridization for each strain that served
as a "fingerprint." The clone pLM-6 did not cross-react with other genera of
anaerobic or aerobic fungi examined (23). The cloned fragment is 675 base
pairs in length and has subsequently been sequenced (A. G. Brownlee, unpublished), revealing an internal complexity of organization with a G + C
content of 35%. The hybridization results (23) indicate a unique arrangement of the repeat in the different isolates of this genus and provide further
evidence for dispersed A + T-rich sequences in strains of *Neocallimastix*.

B. Cloned Coding Sequences

More recently, the cloning of particular genes from *Neocallimastix frontalis*
has been reported (24,25). Using a part of the cellobiohydrolase I (CBHI)

gene of the ascomycete fungus *Trichoderma reesei* as a probe to screen a genomic library of *N. frontalis*, several clones with homology to the probe were identified (25). Subcloning and further analysis, including sequencing, of one of these genomic clones located a putative coding sequence containing small introns with the flanking consensus sequences characteristic of the genes of aerobic fungi (25). Interestingly, the G + C content of the sequenced coding region was 52%, with indications that the flanking regions were quite A + T-rich (R. Durand, personal communication). The retention of DNA sequence homology with a gene from a very distantly related fungus suggests that this approach will be useful for future isolation and characterization of further genes of anaerobic fungi. The *T. reesei* endoglucanase gene (EGLI) is reported to have some homology with *N. frontalis* genomic DNA by dot blot hybridization analysis (24).

A complementary DNA (cDNA) library, made from polyadenylated RNA from cellulose-induced cultures of *N. frontalis*, was screened by differential hybridization with cDNA probes constructed from both cellulose-induced and noninduced polyadenylated RNAs (26). Those clones that hybridized only with the probes derived from cellulose-induced cultures were further screened for hybridization to a portion of the CBHI gene of *T. reesei*. A clone with homology to the CBHI probe was isolated and used to identify a 2.3-Kb RNA transcript, of similar size to the cDNA, present in polyadenylated RNA isolated from cellulose-induced (and to some extent in uninduced) cultures. This evidence suggests that a gene encoding a cellulose-inducible protein (perhaps a *Neocallimastix* cellobiohydrolase), with homology to the *T. reesei* CBHI gene, has been cloned. The transcript was substantially larger, however, than the corresponding messenger RNA identified in aerobic filamentous fungi.

The cDNA for phosphoenolpyruvate carboxykinase from *N. frontalis* has been cloned and sequenced (27). The gene appeared to be transcribed at a high level in cellulose-grown cultures. Codon usage indicated that only 40 codons out of 61 were used, and there was a preference for T in the third position (69% frequency) (27). Three cellulase cDNAs have been isolated from *N. patriciarum* (28). There was no detectable sequence homology between them using Southern hybridization. The cloned cellulases were all expressed in the bacterium *Escherichia coli* using a bacteriophage expression vector (28). A xylanase cDNA (*xynA*) was also isolated from *N. patriciarum* (29). Sequence analysis of *xynA* revealed a bias toward codons ending in T and the use of only 47 of the possible codons, again consistent with the low genomic G + C content. The presence of two repeated regions within the inferred protein sequence suggested that the present gene was derived from an ancestral sequence that was internally duplicated twice. In addition, the authors provided evidence, based on sequence similarities with bacterial xylanase, that *N. patriciarum* may have acquired the gene in the past by horizontal transfer from a rumen procaryote (29).

C. The Ribosomal RNA Genes

The general ease of study and utility of the rRNA genes have made them a focus of study in the author's laboratory, and they will be considered separately in this section.

The rRNA genes of *Neocallimastix* are tandemly arranged in a large cluster (9), as is typical of the organization in other fungi (30). The size of the repeating unit is about 11 Kb (revised from an earlier estimate of 9.4 Kb), on the basis of Southern blots of genomic DNA (see below). Again, this is of the same order as the size of the repeat reported for a range of other groups of fungi (30). The repeat can be separated as a denser band of DNA following cesium chloride density gradient centrifugation, and the $G + C$ content is about 30% (9). This low $G + C$ content is probably the result of very $A + T$-rich spacer regions of DNA flanking the rRNA coding regions; the small ribosomal subunit (18S) RNA gene of *Neocallimastix* has been sequenced (31), and while the $G + C$ content of 41.9% is lower than that of other fungal gene sequences available to date (31), these are few in number. In addition, the results of restriction enzyme digests of genomic DNA, following which the rRNA genes were detected by DNA hybridization (9), as well as direct analysis of the ribosomal spacer region by in vitro DNA amplification (see below), confirm the high $A + T$ content. Experiments using the latter technique give support to the hypothesis that the genes are arranged in the typical eucaryotic fashion, i.e., 18S–5.8S–28S.

The conservation of sequence within the rRNA genes, coupled with variation in the nucleotide sequences separating the genes, has been exploited to assess genetic variation in a number of organisms (32,33,34). Figure 1 indicates how application of this technique to the anaerobic fungi has been used to examine the same question. Using a heterologous DNA probe (the cloned ribosomal RNA gene cluster from the ascomycete fungus *Neurospora crassa* [9]), as in this instance, limits the analysis somewhat: length variation in portions of the rRNA repeat with no homology to the DNA probe (the intergenic regions, for instance) can pass undetected unless they remain linked to the genes that are detected by the probe following restriction enzyme digestion. This may account for the discrepancy in the estimates of the total size of the repeat given earlier (9) and here. Certain restriction fragments produced by the enzymes *Ssp*1 and *Eco*R1 have apparently escaped detection in this analysis (Fig. 1), since the sums of the sizes of the visible fragments for each strain fall short of those visible in the *Hin*D111 and *Bss*H11 digests. Nevertheless, there is evidence from these results for the presence of detectable sequence variation (as so-called restriction fragment length poly-

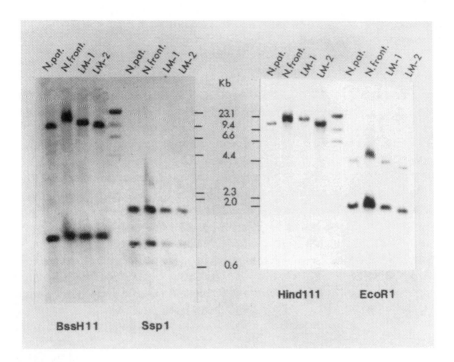

Figure 1 Restriction fragment length variation within the ribosomal repeat of strains of *Neocallimastix*. DNA from *N. patriciarum* (N.pat.), *N. frontalis* (N.front), and Sydney isolates LM-1 and LM-2 was digested with the restriction enzymes indicated and separated by electrophoresis in horizontal agarose gels, and the rRNA fragments were detected after transferal of the DNA to nylon membranes (9). Each group of four strains is separated by DNA size standards (lane 5 on each blot), with sizes (kilobases [Kb]) indicated between the blots.

morphisms [RFLPs]) between morphologically very similar isolates of *Neocallimastix*, including the two described species *N. frontalis* and *N. patriciarum* (note the variation in length of restriction fragments produced by the enzyme *Eco*R1, for example). As mentioned above, the size of the repeat from this analysis is about 11 Kb. The restriction enzyme *Ssp*1 recognizes the hexanucleotide sequence AATATT. This enzyme cleaves the ribosomal repeat frequently (Fig. 1), providing evidence of high A + T content within the repeat (see also Ref. 9), and produces an identical and distinctive pattern of fragments for all the *Neocallimastix* strains shown, as well as others not shown (A. G. Brownlee, unpublished results). Uniflagellate strains of anaerobic fungi tested in the same way yield different results with *Ssp*1.

More recently, a new technique has become available for directly examining the intergenic regions of the repeat (where the majority of the variation is expected to occur) in a convenient and highly sensitive assay involving in vitro amplification of the region with the polymerase chain reaction (PCR) (35). An example of the use of this method for comparing isolates of anaerobic fungi is shown in Fig. 2. Length variation in the internal transcribed

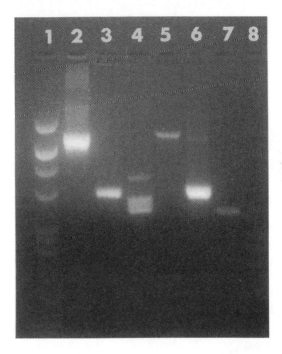

Figure 2 Polymerase chain reaction (PCR) amplification of ribosomal spacer DNA in anaerobic fungi. Polymerase chain reaction products were separated by electrophoresis in 3% Nu-Sieve agarose (FMC BioProducts, USA) gels and detected by staining with ethidium bromide and illumination with ultraviolet light. DNA from *Neocallimastix* strain LM-2 (lanes 2–4) and *Piromyces* strain SM-1 (lanes 5–7) was amplified in a conventional PCR reaction (35,37) with oligonucleotide primers 18SR (rightward from 18S gene) and 28SL (leftward from 28S gene), lanes 2 and 5; 5.8R (rightward from 5.8S gene) and 28SL, lanes 3 and 6; and 5.8L (leftward from 5.8S gene) and 18SR, lanes 4 and 7. Lane 1 is pBR322 plasmid DNA digested with the enzyme Alu1 and serves as a size marker. Primers were designed to be complementary to conserved regions of eucaryotic rRNA genes and have the following sequences (5' to 3'): 18SR, GTTTCCGTAGGTGAACCTGC; 5.8R, GCATCGATGAAGAACGCAGC; 5.8L, the complement of 5.8R; 28SL, ATATGCTTAAGTTCAGCGGGT. Full details will be reported elsewhere.

spacer, separating the 18S and 28S genes, is apparent, and the method confirms the predicted orientation of the respective genes within the cluster (because of the known orientation of the primers used in each reaction). The multiple bands evident in the LM-2 DNA amplified with 5.8L/18SR (Fig. 2, lane 4) are an artifact not usually seen. Restriction enzyme analysis of the amplified fragments (36,37) provides information on RFLPs. As might be predicted, even these small amplified fragments (the 18SR/28SL fragment is about 800 base pairs in length; see Fig. 2, lanes 2 and 5) contain restriction sites for enzymes such as $Ssp1$, indicating their high A + T content (results not shown).

Extensive use of this method has proven extremely useful for analyzing the genetic variability within, and between, the major morphological groups of anaerobic fungi and for providing signatures for individual isolates.

VII. RIBOSOMAL RNA SEQUENCE AND THE MOLECULAR PHYLOGENY OF ANAEROBIC FUNGI

The ribosomal small subunit (18S) RNA gene of *Neocallimastix* has been sequenced using the technique of amplifying the gene with the PCR followed by molecular cloning of the amplified DNA fragment (31,38). The partial sequence of the 18S rRNA molecule itself has been reported for the following species: *N. joyonii, N. frontalis, Piromyces communis,* and *Caecomyces communis* (39,40). This RNA molecule has received widespread attention, because it is highly suited for the construction of molecular phylogenies of organisms (41). By aligning the nucleotide sequences of this highly conserved gene from different organisms, and noting the positions at which the sequences differ, a quantitative estimate can be derived for the "similarity" of the compared organisms. By analyzing these numerical values phylogenetic relationships can be inferred and dendrograms constructed (42,43). In this way it has been demonstrated that the anaerobic fungi are very closely related to each other and that their classification in the order Spizellomycetales of the Chytridiomycota (44) is very strongly supported (31) (Fig. 3). The anaerobic genera described above, for example, share 98% sequence similarity out of 1500 nucleotides examined (39,40), while, in our work, partial sequence analysis of *C. communis* strain NM1 and *P. communis* strain SM1 (representing some 30% of the gene sequence) revealed no detectable sequence differences from each other or from *Neocallimastix* strain LM2 (T. White and A. G. Brownlee, unpublished results). Likewise, when the 18S rRNA sequence of *Neocallimastix* was compared with that obtained for representatives of the Chytridiales (*Chytridium confervae*), the Blastocladiales (*Blastocladiella emersonii*), and the Spizellomycetales (*Spizellomyces acuminatus*), the anaerobe consistently showed association with the Spizellomycetes

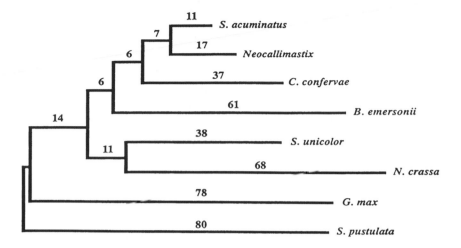

Figure 3 A phylogenetic tree inferred from the aligned 18S rRNA sequences of *Neocallimastix*, the major orders of Chytrids, and other fungi (31). The tree was constructed using the neighbor-joining method (see Ref. 31) and is shown in an unrooted form, with the protist *Stylonychia pustulata* as the outgroup. Only those sequence positions alignable in all 8 species were used in the distance analysis, and the accumulated distances (sequence differences) are indicated on the horizontal branches of the tree. See text for further details.

(31) (Fig. 3). This study was also significant in that it provided strong confirmatory evidence that the chytrids share affinities with the true fungi — the Ascomycetes and the Basidiomycetes — to the exclusion of the protists, or Protoctista, with which they have been grouped in a recent taxonomic scheme (45). Thus it can be seen from Fig. 3 that the ascomycete fungus *N. crassa* and the basidiomycete *Spongipellis unicolor* (the nonflagellar fungi) share a common ancestor (are monophyletic) with the chytrids, to the exclusion of soybean (*Glycine max*) and the protists.

Sequence analysis of the internal transcribed spacer (see Section VI.C) and its adjacent regions from four anaerobic fungi has been used to assess phylogenetic relationships (46). This analysis again grouped the gut fungi, as well as the Spizellomycetales and the Chytridiales, with the Fungi, but could not resolve their immediate relationship. Similarly, the analysis indicated that *Anaeromyces* is a more distant genus, but *Orpinomyces, Neocallimastix,* and *Piromyces* could not be clearly resolved (46).

It seems likely that the enormous information content of nucleotide sequences will play an increasingly important role in the future taxonomy of the anaerobic fungi.

VIII. FUNCTIONAL APPLICATION OF NUCLEOTIDE RESEARCH

The cloning of structural genes and the resultant sequence information have an important contribution to make to the understanding of the biochemistry and physiology of the anaerobic fungi, and of their role in ruminal digestion. As discussed above, molecular studies based on nucleotide sequences can also provide the basis for taxonomic comparisons, and strain identification is greatly facilitated by the very sensitive and reproducible techniques of molecular biology.

Another area of research where the nucleic acids will play an important part is enumeration of the fungi in the rumen. The current methods for estimating fungal numbers in the digestive tract are based on culture or counts of zoospores, sporangia, or thallus-forming units (47–50). The reliability and accuracy of these methods is difficult to judge, although they do provide an adequate measure of relative population changes. DNA probes have been used to follow population numbers of rumen bacteria (51,52), and the results have been quite surprising. The rRNA genes of the fungi are a useful target for DNA-based probes because they are multicopy and there is little sequence variation across anaerobic genera (see above). On the basis of the 18S rRNA gene sequence of *Neocallimastix*, unique regions of sequence have been identified and oligonucleotides have been synthesized (A. G. Brownlee, unpublished results). These can be applied in the analysis of the fungal biomass in the rumen in the same way that the rumen bacteria *Bacteroides succinogenes* and *Lachnospira multiparus* were studied (52). Experiments along these lines are currently in progress.

Much recent evidence points to a wide diversity of types of fungi inhabiting the digestive tract (53–56). With suitable unique DNA probes, individual types of fungi can be identified in digesta samples, and it should be possible to follow the population dynamics. The cloned repetitive sequence pLM-6 described above could be used to follow population numbers of *Neocallimastix* strains, for instance, if the sensitivity is such that cross-hybridization with other elements of the gut contents does not interfere. This aspect has not been investigated as yet, although the probe did not hybridize to DNA from other genera of anaerobic fungi (23). Experiments are under way in the author's laboratory to identify regions of the ribosomal repeat unique to each of the major groups of anaerobic fungi so that specific oligonucleotides can be designed for use as probes. The evidence from the results of PCR amplification of this region (Fig. 2) indicates areas where these might be located.

REFERENCES

1. Taylor JW, Natvig DO. Isolation of fungal DNA. In: Fuller MS, Jaworski A, eds. Zoosporic Fungi in Teaching and Research. Athens, Georgia: Southeastern Publishing Corp., 1987:252–258.

2. Brownlee AG. A rapid DNA isolation procedure applicable to many refractory filamentous fungi. Fungal Genet Newslett 1988; 35:8–9.

3. Paul JH, Myers B. Fluorometric determination of DNA in aquatic microorganisms by use of Hoechst 33258. Appl Environ Microbiol 1982; 43:1393–1399.

4. Phillips MW, Gordon GLR. Growth characteristics on cellobiose of three different anaerobic fungi isolated from the ovine rumen. Appl Environ Microbiol 1989; 55:1695–1702.

5. Skinner DM, Triplett LL. The selective loss of DNA satellites on deproteinization with phenol. Biochem Biophys Res Comm 1967; 28:892–897.

6. Tsai K-P, Calza RE. Enzyme based DNA extraction from zoospores of ruminal fungi. Fungal Genet Newslett 1992; 39:86–88.

7. Timberlake WE. Low repetitive DNA content in *Aspergillus nidulans*. Science 1978; 202:973–975.

8. Kurtzman CP. DNA base sequence complementarity and the definition of fungal taxa. Microbiol Sci 1984; 1:44–48.

9. Brownlee AG. Remarkably AT-rich genomic DNA from the anaerobic fungus *Neocallimastix*. Nucl Acids Res 1989; 17:1327–1335.

10. Billon-Grand G, Fiol JB, Breton A. DNA of some anaerobic rumen fungi: GC content determination. 4th Int Mycol Congr 1990; Abs ID-172/1:p172.

11. Billon-Grand G, Fiol JB, Breton A, et al. DNA of some anaerobic fungi – G + C content determination. FEMS Microbiol Lett 1991; 82:267–270.

12. Fiol JB, Billon-Grand G, Oulhaj Z, et al. Determination du contenu en guanine-cytosine de champignons anaerobies du rumen: *Neocallimastix, Piromonas* et *Sphaeromonas*. Ann Zootech 1992; 41:77–78.

13. Fasman GD, ed. CRC Handbook of Biochemistry and Molecular Biology. 3d ed. Cleveland: CRC Press, 1976.

14. Antequera F, Tamame M, Villanueva JR, Santos T. DNA methylation in the fungi. J Biol Chem 1984; 259:8033–8036.

15. Jupe ER, Magill JM, Magill CW. Stage specific DNA methylation in a fungal plant pathogen. J Bacteriol 1986; 165:420–423.

16. Cano C, Herrera-Estrella L, Ruiz-Herrera J. DNA methylation and polyamines in regulation of development of the fungus *Mucor rouxii*. J Bacteriol 1988; 170: 5946–5948.

17. Jukes T, Bhushan V. Silent nucleotide substitutions and G + C content of some mitochondrial and bacterial genes. J Mol Evol 1986; 24:39–44.

18. Osawa S, Jukes TH, Muta A, et al. Role of directional mutation pressure in the evolution of the eubacterial genetic code. Cold Spring Harbor Symp Quant Biol 1987; 52:777–789.

19. Brownlee AG. What determines the amino acid composition of rumen microbes? Asian-Aust J Anim Sci 1989; 2:441–443.

20. Hyde JE, Kelly SL, Holloway SP, et al. A general approach to isolating *Plasmodium falciparum* genes using non-redundant oligonucleotides inferred from protein sequences of other organisms. Mol Biochem Parasitol 1989; 32:247–262.

21. Szafranski P, Godson GN. Hypersensitive mung bean nuclease cleavage sites in *Plasmodium knowlesi* DNA. Gene 1990; 88:141–147.

22. Taylor JW. Fungal evolutionary biology and mitochondrial DNA. Exp Mycol 1986; 10:259–269.

23. Brownlee AG. A genus-specific, repetitive DNA probe for *Neocallimastix*. In: Nolan JV, Leng RA, Demeyer DI, eds. The Roles of Protozoa and Fungi in Ruminant Digestion. Armidale, Australia: Penambul Books, 1989:251–253.

24. Reymond P, Durand R, Hébraud M, Fèvre M. Molecular cloning of genes from the rumen anaerobic fungus *Neocallimastix frontalis*. Expression during hydrolase induction. FEMS Microbiol Lett 1991; 77:107–112.

25. Durand R, Reymond P, Fèvre M. Molecular cloning of a gene from the rumen anaerobic fungus *Neocallimastix frontalis* showing homology to the cellobiohydrolase gene from *Trichoderma reesei*. 4th Int Mycol Congr 1990; Abs ID-181/3:p181.

26. Reymond P, Durand R, Fevre M. Molecular cloning of cDNAs regulated during induction of cell wall degrading enzymes in *Neocallimastix frontalis*. 4th Int Mycol Congr 1990; Abs ID-208/4:p208.

27. Reymond P, Geourjon C, Roux B, et al. Sequence of the phosphoenolpyruvate carboxykinase–encoding cDNA from the rumen anaerobic fungus *Neocallimastix frontalis*: comparison of the amino acid sequence with animals and yeast. Gene 1992; 110:57–63.

28. Xue G-P, Orpin CG, Gobius KS, Aylward JH, Simpson GD. Cloning and expression of multiple cellulase cDNAs from the anaerobic rumen fungus *Neocallimastix patriciarum* in *Escherichia coli*. J Gen Microbiol 1992; 138:1413–1420.

29. Gilbert HJ, Hazlewood GP, Laurie JI, et al. Homologous catalytic domains in a rumen fungal xylanase: evidence for gene duplication and prokaryotic origin. Mol Microbiol 1992; 6:2065–2072.

30. Garber RC, Turgeon BG, Selker EU, Yoder OC. Organisation of ribosomal RNA genes in the fungus *Cochliobolus heterostrophus*. Curr Genet 1988; 14:573–582.

31. Bowman B, Taylor JW, Brownlee AG, et al. Molecular evolution of the fungi: relationship of the basidiomycetes, ascomycetes and chytridiomycetes. Molec Biol Evol 1992; 9:285–296.

32. Wostemeyer J. Strain-dependent variation in ribosomal DNA arrangement in *Absidia glauca*. Eur J Biochem 1985; 146:443–448.

33. Magee BB, D'Souza TM, Magee PT. Strain and species identification by restriction fragment length polymorphisms in the ribosomal DNA repeat of Candida species. J Bacteriol 1987; 169:1639–1643.

34. Klassen GR, McNabb SA, Dick MW. Comparison of physical maps of ribosomal DNA repeating units in *Pythium, Phytophthora* and *Apodachlya*. J Gen Microbiol 1987; 133:2953–2959.

35. Saiki RK, Gelfand DH, Stoffel S, et al. Primer-directed enzymatic amplification with a thermostable DNA polymerase. Science 1988; 239:487–491.

36. White TJ, Bruns TD, Lee SB, Taylor JW. Amplification and direct sequencing of fungal ribosomal RNA genes for phylogenetics. In: Innis N, Gelfand D, Sninsky J, White T, eds. PCR-Protocols: A Guide to Methods and Applications. New York: Academic Press, 1990:315–322.

37. Gardes M, White TJ, Fortin JA, et al. Identification of indigenous and introduced symbiotic fungi in ectomycorrhizae by amplification of nuclear and mitochondrial ribosomal DNA. Can J Bot 1991; 69:180–190.

38. Brownlee AG. The systematic placement of rumen anaerobic fungi based on small ribosomal RNA sequence. 4th Int Mycol Congr 1990; Abs IG-319/3:p319.

39. Doré J, Stahl DA. Phylogeny of anaerobic rumen phycomycetes based on small ribosomal subunit RNA-sequence comparison. 4th Int Mycol Congr 1990; Abs IIC-117/3:p117.

40. Doré J, Stahl DA. Phylogeny of anaerobic rumen Chytridiomycetes inferred from small subunit ribosomal RNA sequence comparisons. Can J Bot 1991; 1964–1971.

41. Sogin M. Evolution of eukaryotic microorganisms and their small subunit ribosomal RNAs. Am Zool 1989; 29:487–499.

42. Lane DJ, Pace B, Olsen GJ, Stahl DA, Sogin ML, Pace NR. Rapid determination of 16S ribosomal RNA sequences for phylogenetic analyses. Proc Natl Acad Sci USA 1985; 82:6955–6959.

43. Woese CR. Bacterial evolution. Microbiol Rev 1987; 51:221–271.

44. Heath IB, Bauchop T, Skipp RA. Assignment of the rumen anaerobe *Neocallimastix frontalis* to the Spizellomycetales (Chytridiomycetes) on the basis of its polyflagellate zoospore ultrastructure. Can J Bot 1983; 671:295–307.

45. Margulis L, Schwartz KV. Five Kingdoms. New York: Freeman, 1988.

46. Li J, Heath IB. The phylogenetic relationships of the anaerobic chytridiomycetous gut fungi (Neocallimasticaceae) and the Chytridiomycota. I. Cladistic analysis of rRNA sequences. Can J Bot 1992; 70:1738–1746.

47. Orpin CG. On the induction of zoosporogenesis in the rumen phycomycetes *Neocallimastix frontalis, Piromonas communis* and *Sphaeromonas communis*. J Gen Microbiol 1977; 101:181–189.

48. Joblin KN. Isolation, enumeration, and maintenance of rumen anaerobic fungi in roll tubes. Appl Environ Microbiol 1981; 42:1119–1122.

49. Ushida K, Tanaka H, Kojima Y. A simple *in situ* method for estimating fungal population size in the rumen. Lett Appl Microbiol 1989; 9:109–111.

50. Theodorou MK, Gill M, King-Spooner C, Beever D. Enumeration of anaerobic Chytridiomycetes as thallus-forming units: novel method for quantification of fibrolytic fungal populations from the digestive tract ecosystem. Appl Environ Microbiol 1990; 56:1073–1078.

51. Attwood GT, Lockington RA, Xue G-P, Brooker JD. Use of a unique gene sequence as a probe to enumerate a strain of *Bacteroides ruminicola* introduced into the rumen. Appl Environ Microbiol 1988; 54:534–539.

52. Stahl DA, Flesher B, Mansfield HR, Montgomery L. Use of phylogenetically based hybridization probes for studies of ruminal microbial ecology. Appl Environ Microbiol 1988; 54:1079–1084.

53. Heath IB. Gut fungi. Trends Ecol Evol 1988; 3:167–171.

54. Barr DJS, Kudo H, Jakober KD, Cheng K-J. Morphology and development of rumen fungi: *Neocallimastix* sp., *Piromyces communis*, and *Orpinomyces bovis* gen. nov., sp. nov. Can J Bot 1989; 67:2815–2824.

55. Breton A, Bernalier A, Dusser M, et al. *Anaeromyces mucronatus* nov. gen., nov. sp. A new strictly anaerobic rumen fungus with polycentric thallus. FEMS Microbiol Lett 1990; 70:177–182.

56. Ho YW, Bauchop T, Abdullah N, Jalaludin S. *Ruminomyces elegans* gen. et sp. nov., a polycentric anaerobic rumen fungus from cattle. Mycotaxon 1990; 38:397–405.

10

Differential and Integral Equations and Their Application in Quantifying the Fungal Population in the Rumen

JAMES FRANCE

AFRC Institute of Grassland and Environmental Research
Okehampton, England

MICHAEL K. THEODOROU

Institute of Grassland and Environmental Research
Plas Gogerddan, Aberystwyth
Dyfed, Wales

I. INTRODUCTION

In the sciences, mathematical models are developed to aid in the understanding of physical, chemical, and biological phenomena. These models often yield equations containing derivatives of unknown functions. Such equations are referred to as differential equations. Differential equations are central to the sciences: it is often claimed that Sir Isaac Newton's great discovery was that they provide the key to the "system of the world."

Differential equations arise within biology in the construction of dynamic, deterministic, mechanistic models. There is a mathematically standard way of representing such models called the rate:state formalism. The system under investigation is defined at time t by q state variables: $X_1, X_2, ..., X_q$. These variables represent properties or attributes of the system, such as microbial mass or population, quantity of substrate, etc. The model then comprises q first order differential equataions that describe how the state variables change with respect to time:

$$dX_i/dt = f_i(X_1, X_2, ..., X_q; S); i = 1, 2, ..., q \qquad (1.1)$$

where S denotes a set of parameters, and the function f_i gives the rate of change of the state variable X_i. The function f_i comprises terms that represent the rates of processes (with dimensions of state variable per unit time), and these rates can be calculated from the values of the state variables alone, with, of course, the values of any parameters and constants. In this type of mathematical modeling, the differential equations are formed through direct application of the laws of science (e.g., the law of mass conservation, the first law of thermodynamics) or by application of a continuity equation derived from more fundamental scientific laws.

Sometimes it is convenient to express a differential equation as an integral equation; for example, equations (1.1) may be written:

$$X_i = X_i(0) + {}_0\!\int^t f_i(X_1, X_2, ..., X_q; S)dt; i = 1, 2, ..., q \tag{1.2}$$

where $X_i(0)$ denotes the initial (zero time) value of X_i. Integral equations arise not only as the converse of differential equations, but also in their own right. For example, the response of a system sometimes depends not just on the state of the system per se but also on the form of the input. Input P and output U might then be related by the convolution (or Faltung) integral:

$$U(t) = {}_0\!\int^t P(x)W(t - x)dx = P(t)*W(t) \tag{1.3}$$

where x is a dummy variable ranging over the time interval zero to the present time t during which the input has occurred, and W is a weighting function that weights past values of the input to the present value of the output. The symbol * denotes the convolution operator.

Integral equations are much less common in biology than differential equations (though they occur frequently as convolution integrals in areas such as tracer kinetics). One of the best-known examples in biology is the following integro-differential equation:

$$(1/y)dy/dt = a + by + {}_c\!\int^t K(t,x)y(x)dx \tag{1.4}$$

in which y denotes population size, the kernel K represents the coefficient of heredity, x is a dummy variable, and a, b, and c are parameters. This equation was suggested by Volterra in the early 1900s as a device to take account of hereditary influences in the study of population growth. In certain cases, the equation involved a convolution. In this chapter, the application of differential and integral equations is explored in the context of quantifying the fungal population in the rumen ecosystem. Two methods of estimation are described; one based on the life cycle of anaerobic fungi and employing differential equations, and the other based on incubating particles of different size in rumen liquid and employing integral equations.

II. A LIFE-CYCLE MODEL

A. Biological Background

An understanding of the fungal life cycle and its role within the rumen is necessary to the presentation of this model. The rumen is an open (continuous) ecosystem in which the continuity of dietary inputs, passage of liquid and solids, and absorption of fermentation end products ensures a relatively unchanging environment (stable state). Under these conditions, and in the absence of any dramatic dietary perturbation, the population of fungi in the rumen remains relatively constant (1). In order for this to occur, growth of fungal biomass in the rumen must balance the loss of fungi by death and/or passage from the rumen. This is unlike the growth characteristics of anaerobic fungi in laboratory batch cultures, where rapidly changing conditions are not tolerated and fungi fail to survive for more than a few days without repeated subculture (1,2). Apart from the polycentric forms (3), all anaerobic fungi have a classical monocentric life cycle, in which free-swimming zoospores alternate with particle-attached (substrate-associated) thalli and zoosporulation is followed by death of the remaining thallus (4,5). Therefore, for these fungi to remain viable in the rumen, both zoospores and thalli must coexist in a state of dynamic equilibrium. Thus, it follows for the monocentric fungi that a nontrivial relationship exists between the numbers of zoospores and of thalli in the rumen ecosystem. Although we recognize that polycentric fungi may be of major significance in the rumen, a paucity of data regarding their life histories and mode of zoospore production prevents their detailed consideration here.

Zoospores are small in comparison with thalli and their existence more limited in time (4), and thus it is reasonable to assume that most fungal biomass in the rumen is in the form of particle-attached thalli. These thalli represent the functional unit of the fungus with respect to plant cell-wall degradation, and their enumeration and biomass estimation are of paramount importance when ascribing a role to anaerobic fungi in the rumen. However, because of the difficulty of quantifying particle-attached fungal biomass in a mixed-population ecosystem, direct enumeration of the thallus population has not yet been accomplished effectively. Several procedures are available for enumeration of zoospores. These involve direct microscopic counts of zoospores in rumen liquid or indirect viable counts of colonies developing from zoospores in agar-containing roll tubes (6,7). A lesser-used technique, which can be used to compare the development of thallus populations from zoospores, is to count the number of zoosporangia associated with leaf blades or impregnated agar strips after in vitro or in sacco incubations in rumen liquid (8,9). An enumeration procedure based on the most-probable-number technique is also available. This technique does not distinguish between stages of the life cycle, but it can provide a minimal enumeration of the entire fungal population (1).

While several procedures can be used to accomplish some degree of quantification of the fungal population in the rumen, none is wholly suitable for enumeration of fungal thalli or estimation of biomass. However, owing to the dynamic equilibrium between zoospores and thalli in the rumen, knowledge of the fungal life cycle can be used to determine the numbers of fungal thalli associated with digesta solids from the concentration of zoospores in rumen liquid. A mathematical model to estimate fungal thalli, which utilizes knowledge of the life cycle of monocentric fungi obtained from in vitro and in vivo investigations, together with in vivo information on rumen volumes and the flow of liquid and digesta from the rumen, is presented herein. As it is the primary intention in this chapter to illustrate the application of differential and integral equations to the rumen ecosystem, a more detailed account of the model and its application is left to France et al. (10).

B. Mathematical Formulation

In the model, the life cycle of anaerobic fungi (Fig. 1) is partitioned so that the rumen population, X, is considered in terms of three mutually exclusive subpopulations X_1, X_2, and X_3. X_1 denotes the number of free-swimming zoospores (those in the first phase of their life cycle), X_2 the number of particle-attached thalli undergoing vegetative growth before septation (the second phase of the cycle), and X_3 the number of mature, particle-attached thalli between septation and zoosporulation (the third phase). It is assumed that the time spent in each phase of the life cycle (by those individuals that will enter the next phase) follows an exponential distribution with respective known means λ_1, λ_2, and λ_3 (h). On completion of the life cycle after zoosporogenesis, Y motile zoospores per zoosporangium are released back into the liquid medium.

A schematic representation of the life cycle and ruminal events assumed in the model is shown in Fig. 2. The state variables X_1, X_2, and X_3 are depicted as compartments (boxes) within the rumen, and the fluxes between compartments and out of the system as arrowed lines. Only those fungi in the first phase of their life cycle are assumed to leave the rumen with the passage of liquid digesta to the lower gut, and only those in the second and third phases with the passage of particulate digesta. To effect a solution, all passage is taken to be a first-order process with a rate constant k_1 for liquid and k_2 for particulate matter (both per h). Similarly, death in each phase is assumed first-order with respective rate constants d_1, d_2, d_3 (all per h). The rate constant pertaining to each flux is shown against the corresponding

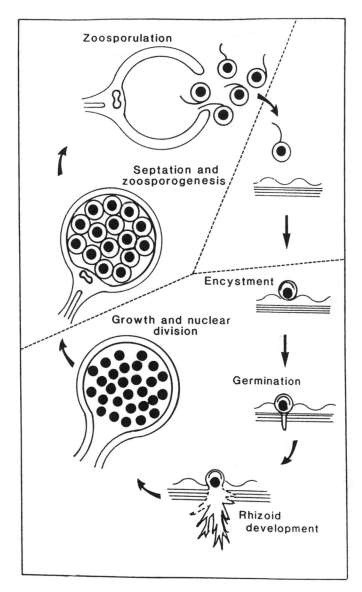

Figure 1 Diagrammatic representation of the life cycle of monocentric rumen fungi. The cycle was partitioned into three sections representing populations of zoospores, immature thalli, and mature thalli, respectively (as indicated by the broken lines), for translation into a mathematical model. (Reproduced by permission of Academic Press.)

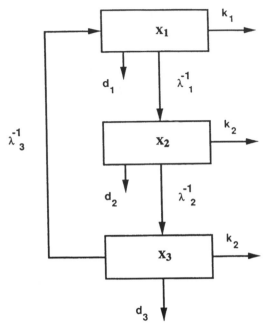

Figure 2 Compartmental scheme for estimating the population of anaerobic fungi in the rumen ecosystem. The three compartments X_1, X_2, and X_3 represent populations of zoospores, immature thalli, and mature thalli, respectively. The arrowed lines represent fluxes, and the rate constant pertaining to each flux is shown against the corresponding arrowed line. (Reproduced by permission of Academic Press.)

arrowed line. Applying the law of mass conservation, the differential equations describing the dynamics of this system are:

$$dX_1/dt = \lambda_3^{-1}YX_3 - (k_1 + d_1 + \lambda_1^{-1})X_1 \qquad (2.1)$$

$$dX_2/dt = \lambda_1^{-1}X_1 - (k_2 + d_2 + \lambda_2^{-1})X_2 \qquad (2.2)$$

$$dX_3/dt = \lambda_2^{-1}X_2 - (k_2 + d_3 + \lambda_3^{-1})X_3 \qquad (2.3)$$

These differential equations satisfy the rate:state formalism (cf. equations 1.1). Under the assumption of steady-state conditions in the rumen, equations (2.1)–(2.3) equate to zero. For a nontrivial steady-state solution, the rate constants and yield factor must satisfy the following condition, namely:

$$(k_1 + d_1 + \lambda_1^{-1})(k_2 + d_2 + \lambda_2^{-1})(k_2 + d_3 + \lambda_3^{-1})$$
$$- \lambda_1^{-1}\lambda_2^{-1}\lambda_3^{-1}Y = 0 \qquad (2.4)$$

Numerical solutions to equation (2.4) for the parameters d_1, d_2, d_3, k_1, k_2, λ_1, λ_2, λ_3, and Y (all > 0) can be obtained by imposing an appropriate (representative) range on each of these parameters and solving iteratively using the Newton–Raphson method. The motile population X_1 may be calculated from the concentration of zoospores in rumen liquid, C_{X1} (zoospores/ml), and the rumen liquid volume, V (ml):

$$X_1 = C_{X1}V \tag{2.5}$$

Appropriate values for the parameters d_1, d_2, d_3, k_1, k_2, λ_1, λ_2, λ_3, and Y having been determined by solving (2.4), the thallus populations X_2 and X_3 can then be evaluated using equations (2.1) and (2.2) equated to zero as follows:

$$X_2 = \lambda_1^{-1}X_1/(k_2 + d_2 + \lambda_2^{-1}) \tag{2.6}$$

$$X_3 = (k_1 + d_1 + \lambda_1^{-1})X_1/(\lambda_3^{-1}Y) \tag{2.7}$$

The total number of anaerobic fungi in the rumen is given by the sum:

$$X = X_1 + X_2 + X_3 \tag{2.8}$$

Therefore, the total population of anaerobic fungi, together with the number in each phase of the life cycle, can be estimated by solving nonlinear equations (2.4)–(2.8) numerically subject to appropriate ranges.

C. Numerical Solution

Ranges on the parameters can be constructed by recourse to the published literature, as follows. The life cycles of several monocentric rumen fungi have been described (e.g., *Neocallimastix patriciarum, N. frontalis, N. hurleyensis, Piromonas communis, Sphaeromonas communis,* and *Caecomyces equi*) (11–16). A number of studies on aspects of the life cycles of several monocentric rumen fungi have also been reported (4,6,7,17,18,19). Nonetheless, there is insufficient quantitative information in the literature to ascribe accurate ranges, let alone exact values, to all the life-cycle parameters for any individual species of rumen fungus. Therefore, parameter ranges appropriate to rumen fungi generally have to be adopted. On the basis of an extensive review of the literature (10), we suggest the following ranges for the life-cycle parameters: $0.25 \leqslant \lambda_1 \leqslant 4$ h, $6 \leqslant \lambda_2 \leqslant 24$ h, $0.25 \leqslant \lambda_3 \leqslant 18$ h, and $10 \leqslant Y \leqslant 120$.

Information on the fractional rate parameters is uneven. Data in the literature on which to base ranges for the death rates are very sparse and appear to be confined to Theodorou et al. (1). These data imply the ranges: $0.009 \leqslant d_1 \leqslant 0.015$ per h, $0 \leqslant d_2 \leqslant 0.14$ per h, and $0 \leqslant d_3 \leqslant 0.14$ per h. How-

ever, the literature on ruminal passage rates for liquid and solids is much
more extensive, and is summarized by Dhanoa et al. (20). This analysis gives
the ranges: $0.041 \leqslant k_1 \leqslant 0.12$ per h, and $0.020 \leqslant k_2 \leqslant 0.051$ per h.

Solving equations (2.4)–(2.8) numerically subject to these bounds using
the Newton–Raphson method with X_1 set at unity yields $X_2 = 1.28$ and
$X_3 = 0.4$. This solution suggests that in a population of anaerobic fungi,
the number of immature and mature particle-attached thalli is about 70%
greater than the population of free-swimming zoospores. An alternative
solution, obtained by imposing tighter restrictions on the life-cycle para-
meters (i.e., $0.15 \leqslant \lambda_1 \leqslant 1$, $18 \leqslant \lambda_2 \leqslant 24$, $3 \leqslant \lambda_3 \leqslant 6$ and $40 \leqslant Y \leqslant 80$; see Ref.
10), yields $X_2 = 0.46$ and $X_3 = 0.081$. In deriving the solution, as informa-
tion is so sparse, no restrictions were imposed on the death rates other than
that they lay within the same range as the other rate constants (i.e., $0 \leqslant d_1$,
d_2, $d_3 \leqslant 1/0.15$). This solution indicates that the total particle-attached population is approximately 50% of that of the zoospores. Theodorou et al.
(1) describe an experiment in which four British Friesian steers (average live
weight 223 kg) were fed ad libitum on chopped perennial ryegrass hay supplemented with concentrates, and the volume and zoospore concentration
of their rumens were measured. Applying these solutions to that experiment
gives values of 7.5×10^8 for the free-swimming zoospore population in the
rumen and $4.0–12.6 \times 10^8$ for the number of particle-attached thalli, of
which $0.6–3.0 \times 10^8$ are mature. These estimates are broadly consistent with
the minimal ruminal enumeration of $0.9–8.4 \times 10^8$ thallus-forming units
(zoospores plus thalli) reported for the same animals (1).

III. PARTICLE-SIZE MODEL

A. Biological Background

The life-cycle model presented above contains 11 parameters [see equations
(2.4)–(2.8)]. Although these could be determined experimentally in the same
laboratory, the amount of work involved would be prohibitive, and thus
we are constrained to obtain the majority of parameter values from the literature. Even if the assumptions on which the model is based are correct, it
must be emphasized that values for parameters were obtained in different
laboratories using a variety of techniques. This strategy represents a weak-
ness in the life-cycle model, because the research from which values were
taken was not conducted with a view to satisfying model requirements. The
particle-size model to be described is based on fewer assumptions and fewer
parameters. It illustrates a less intensive approach whereby parameter values are generated more readily from experimental observations. The model
is developed from the premise that the number of colonized particles in the
rumen is always less than the number of fungal thalli because each particle

can contain more than one fungal thallus. If it were possible to fragment colonized particles, so that each fragment contained only one fungal thallus, enumeration of the thallus population by counting colonized particles would be comparatively easy. Clearly, it is not possible to achieve this by direct experimentation, although fragmentation can be achieved indirectly by extrapolating mathematically to an infinitesimally small particle size. With infinitesimally small particles, the rate of change in the number of colonized particles can be equated with the rate of attachment of zoospores (i.e., creation of thalli).

The proportion of particles of different size colonized (and hence the rate of change in the number colonized) can be obtained quite simply as follows. A range of particles of known size are suspended in a number of polyester bags in the rumen. At intervals after introduction, bags containing particles of different sizes are removed from the rumen and subsamples are taken for microbiological analysis. Spread-plates are prepared from each of the subsamples using an antibiotic-containing agar medium in petri dishes in an anaerobic chamber. After incubation, these plates are observed under a dissecting microscope to determine the proportion of particles colonized by anaerobic fungi. We have yet to undertake this in sacco study, although the methodology for determining the colonization of particles has been developed in vitro, as illustrated in Fig. 3.

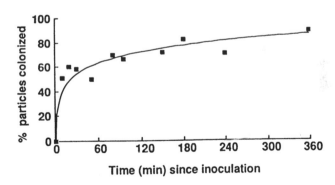

Figure 3   Colonization of particles of wheat straw by *Neocallimastix hurleyensis* in vitro. Particles were removed from batch cultures at intervals after inoculation with zoospores. The number of particles colonized by one or more fungal zoospores was determined using an anaerobic spread-plate procedure. This experimental approach could be used in vivo, by suspending feed particles of different size in polyester bags in the rumen, to obtain parameter values required for the particle-size model. We acknowledge D. Davies and A. P. J. Trinci for these previously unpublished results.

These profiles on the colonization of particles of different size are necessary in order to produce a three-dimensional plot (incubation time versus proportion of particles colonized versus particle size) from which a surface response function can be obtained. The response function can then be used to extrapolate to a limiting condition where each particles is small enough to accommodate no more than one fungal thallus. The slope of the curve of particle colonization against time in this plane of the response function represents the rate of attachment of zoospores and thus the rate of creation of fungal thalli. With this information and from a knowledge of the intake and ruminal turnover time of the particulate matter, the number of thalli in the rumen can be determined.

B. Mathematical Formulation

For ease of exposition, length here is used as an index of particle size. Consider a quantity of forage particles of equal length L (mm) incubated in sacco in rumen liquid at time t = 0 h. The concentration of particulate matter under incubation is assumed to be representative of its concentration in the rumen proper. Let the function $\phi(t, L)$ denote the total number, per gram dry matter (DM) of incubated particulate material, of these particles colonized by motile zoospores by time t after incubation. Assuming that if the length of particles L was selected to be sufficiently small, so that no more than one zoospore could colonize a particle, then the number of associated particle-attached thalli at time t after incubation may be represented by the limit as L tends to zero of function $\phi(t, L)$. Therefore, the rate of particle attachment by the fungi, $\Phi(t)$ (number of thalli/g DM particulate material incubated/h), can be expressed as the rate of change of this limit with respect to time:

$$\Phi(t) = \partial \lim_{L \to 0} [\phi(t,L)] / \partial t \tag{3.1}$$

The number of particle-attached thalli in the rumen can be found from the rate of attachment and the forage intake. Let I(t) denote the rate of intake by the ruminant animal (g DM particulate material/h), then the total number of particle-attached thalli in the rumen X(t) at time t (> T) is given by the double-integral equation:

$$X(t) = \int_{y=t-T}^{t} I(t-y) \int_{x=0}^{y} \Phi(x)dxdy \tag{3.2}$$

where T (h), a constant, is the ruminal turnover time of the thalli. Some simplifying assumptions are necessary to effect a solution. It is assumed, first, that the rumen is in steady state and so I(t) is the same as that for particulate matter. Thus:

$$T = Q/I \tag{3.3}$$

where Q (g DM) is the size of the particulate pool in steady state. Q can be determined directly by rumen emptying or indirectly using markers and polyester bags to estimate the rates of passage and digestion, respectively. Equation (3.2) now simplifies to:

$$X = I \int_{y=0}^{Q/I} \int_{x=0}^{y} \Phi(x)dxdy \tag{3.4}$$

A functional form for Φ is required to solve equation (3.4). On the basis of preliminary observations made at our laboratory (M. K. Theodorou, D. Davies, and J. France, unpublished results) and Fig. 3, it is further postulated that function $\phi(t, L)$, the number of particles colonized by motile zoospores by time t after incubation, is a Mitscherlich equation. The Mitscherlich describes diminishing-returns behavior with asymptote A and rate parameter B. Here, A and B are assumed to depend on the length of particle incubated:

$$\phi(t,L) = A(1 - e^{-Bt}) \tag{3.5}$$

$$A = a_0/(1 + a_1 L^a) \tag{3.6}$$

$$B = b_0 + b_1 L^b \tag{3.7}$$

The a's and b's in equations (3.5)–(3.7) denote constant parameters. Using these equations in equation (3.1) then substituting for Φ in equation (3.4) yields:

$$X = I \int_{0}^{Q/I} \int_{0}^{y} a_0 b_0 e^{-b_0 x} dxdy \tag{3.8a}$$

$$= I \int_{0}^{Q/I} a_0(1 - e^{-b_0 y})dy \tag{3.8b}$$

$$= a_0 Q - a_0 I(1 - e^{-b_0 Q/I})/b_0 \tag{3.8c}$$

Therefore, the quantity of fungal thalli can be determined using the formula given by equation (3.8c), if the intake and rumen content of particulate matter are measured, and the parameters a_0 and b_0 estimated from serial incubation data using particles of different lengths. No numerical solutions to the model are currently available, as the necessary experiments have yet to be undertaken. Hence, the number of different particle lengths required to define the surface response function remains to be determined.

IV. CONCLUSIONS

The rumen fungi are highly fibrolytic microorganisms, producing a wide range of cell-bound and cell-free glycolytic, cellulolytic, and hemicellulolytic enzymes (21,22). In ruminants fed on fibrous diets, a substantial proportion of the plant fragments that enter the rumen is rapidly and extensively colonized by large populations of anaerobic fungi (17). However, the precise role and overall contribution of rumen fungi to the fermentation of plant biomass in the rumen have yet to be fully elucidated.

In this chapter, we have made a case for indirect enumeration of fungal thalli by using mathematical models based on differential or integral equations. At the very least, we have shown that the alternate life cycle of rumen fungi, with their free-swimming dispersal stage and particle-attached vegetative and reproductive stages, is amenable to representation by differential equations. Of the two presented, the life-cycle model has perhaps the greater theoretical appeal, but the particle-size model may be more practicable in that fewer parameters are needed and these are amenable to determination by established experimental techniques within a single laboratory. Although both models are highly speculative at present, they should be considered in the context of what else is known about the quantity and activity of fungal biomass in the rumen. In model construction we have focused our attention on the fungal thallus as a functional unit in plant cell-wall degradation in the rumen. In so doing, we are led to conclude that the most important limitation to experimental research in this area is the restricted opportunity for quantifying fungal thalli by either direct or indirect observation. Although more information is required on the parameters of the models, the general validity of our theoretical approach can be tested only by comparison with other methods. It is therefore necessary to develop new direct and indirect procedures (perhaps involving molecular probes) for enumerating populations of zoospores and thalli in the rumen. For further developments to occur with respect to the models described, it will be necessary to conduct experiments to solve the particle-size model and to determine the biomass associated with an average fungal thallus. Consideration is being given to such experiments in our laboratory at the present time. Solutions to the life-cycle model suggest that zoospore and thallus populations in the rumen are of the same order of magnitude. The published literature also indicates that zoospore and protozan populations in the rumen are numerically similar. If we make a reasonable assumption that the weight of an average fungal thallus is equal to the weight of an average protozoan, it follows that the amount and activity of fungal biomass in the rumen, with respect to plant cell-wall degradation, is likely to be significant.

The reader is referred to the excellent monograph by Czerkawski (23) for further applications of mathematical modelling to rumen studies generally.

ACKNOWLEDGMENT

Dr. J. H. M. Thornley is thanked for critically reading the manuscript.

REFERENCES

1. Theodorou MK, Gill M, King-Spooner C, Beever DE. Enumeration of anaerobic chytridiomycetes as thallus forming units: novel method for quantification of fibrolytic fungal populations from the digestive tract ecosystem. Appl Environ Microbiol 1990; 56:1073–1978.
2. Milne A, Theodorou MK, Jordan MGC, et al. Survival of anaerobic fungi in faeces, in saliva, and in pure culture. Exp Mycol 1989; 13:27–37.
3. Borneman WS, Akin DE, Ljundahl LG. Fermentation products and plant cell wall–degrading enzymes produced by monocentric and polycentric anaerobic ruminal fungi. Appl Environ Microbiol 1989; 55:1066–1073.
4. Lowe SE, Griffith GW, Milne A, et al. The life cycle and growth kinetics of an anaerobic rumen fungus. J Gen Microbiol 1987; 133:1815–1827.
5. Trinci APJ, Lowe SE, Theodorou MK. Growth and survival of rumen fungi. BioSystems 1987; 21:357–363.
6. Orpin CG. Studies on the rumen flagellate *Neocallimastix frontalis*. J Gen Microbiol 1975; 91:249–262.
7. Joblin KN. Isolation, enumeration and maintenance of rumen anaerobic fungi in roll tubes. Appl Environ Microbiol 1981; 42:1119–1122.
8. Akin DE, Gordon GLR, Hogan JP. Rumen bacterial and fungal degradation of *Digitaria pentzii* grown with and without sulphur. Appl Environ Microbiol 1983; 46:738–748.
9. Ushida K, Tanada H, Kojima Y. A simple in situ method for estimating fungal populations in the rumen. Lett Appl Microbiol 1989; 9:109–111.
10. France J, Theodorou MK, Davies D. Use of zoospore concentrations and life cycle parameters in determining the population of anaerobic fungi in the rumen ecosystem. J Theoret Biol 1990; 147:413–422.
11. Gold JJ, Heath IB, Bauchop T. Ultrastructural description of a new chytrid genus of caecum anaerobe, *Caecomyces equi* gen. nov., sp. nov., assigned to the Neocallimasticaceae. BioSystems 1988; 21:403–415.
12. Heath IB, Bauchop T, Skipp RA. Assignment of the rumen anaerobe *Neocallimastix frontalis* to the Spizellomycetales (Chytridiomycetes) on the basis of its polyflagellate zoospore ultrastructure. Can J Bot 1983; 61:295–307.

13. Orpin CG. Studies on the rumen flagellate *Sphaeromonas communis*. J Gen Microbiol 1976; 94:270–280.
14. Orpin CG. The rumen flagellate *Piromonas communis*: its life-history and invasion of plant material in the rumen. J Gen Microbiol 1977; 99:107–117.
15. Orpin CG, Munn EA. *Neocallimastix patriciarum* sp. nov., a new member of the Neocallimasticaceae inhabiting the rumen of sheep. Trans Br Mycol Soc 1986; 86:178–181.
16. Webb J, Theodorou MK. *Neocallimastix hurleyensis* sp. nov., an anaerobic fungus from the ovine rumen. Can J Bot 1991; 69:1220–1224.
17. Bauchop T. Rumen anaerobic fungi of cattle and sheep. Appl Environ Microbiol 1979; 38:148–158.
18. Bauchop T. The anaerobic fungi in rumen fibre digestion. Agric Environ 1981; 6:339–348.
19. Orpin CG. The rumen flagellates *Callimastix frontalis* and *Monas communis*—zoospores of Phycomycete fungi. J Appl Bacteriol 1974; 37:ix–x.
20. Dhanoa MS, France J, Siddons RC. On using a double-exponential model for describing faecal marker concentration curves. J Theoret Biol 1989; 141:247–257.
21. Lowe SE, Theodorou MK, Trinci APJ. Cellulases and xylanase of an anaerobic rumen fungus grown on wheat straw, wheat straw holocellulose, cellulose, and xylan. Appl Environ Microbiol 1987; 53:1216–1223.
22. Williams AG, Orpin CG. Glycoside hydrolase enzymes present in the zoospore and vegetative growth stages of the rumen fungi *Neocallimastix patriciarum*, *Piromonas communis*, and an unidentified isolate grown on a range of carbohydrates. Can J Microbiol 1987; 33:427–434.
23. Czerkawski JW. An Introduction to Rumen Studies. Oxford: Pergamon Press, 1986.

11

Anaerobic Fungi: Future Perspectives

DOUGLAS O. MOUNTFORT

Cawthron Institute
Nelson, New Zealand

I. SIGNIFICANCE AND DISTRIBUTION OF ANAEROBIC FUNGI IN ANAEROBIC ECOSYSTEMS

A. Role in Fiber Degradation

There now appears little doubt that obligate anaerobic fungi are an important component of the microbiota of the rumen of herbivores. Yet it is perplexing that so much time has elapsed before these organisms attracted the widespread interest of researchers. Physiologically they can be grouped with those species of bacteria that participate in the digestion and fermentation of plant material in the rumen. Their mode of growth, involving hyphal extension and possession of powerful xylanolytic and cellulolytic activities (1–9), suggests that they have a special role in digestion of the cellulosic and hemicellulosic components of plant fiber. This capability has now been sufficiently recognized that techniques are being developed for cloning the cellulase genes of anaerobic fungi in other microorganisms, with the purpose of improving rumen function (10–13). Because of their fibrolytic activity anaerobic fungi are now being considered for use in anaerobic digestors, and their H_2-producing capability has been exploited in processes that combine biological H_2-producing reactions with palladium-catalyzed hydrogenations (14). Nevertheless, the role of the anaerobic fungi in the degradation of some

271

plant components such as lignin remains unclear. While several enzymes that cleave the aromatics from the hemicellulosic components of plant cell-walls have been characterized (15,16,17), there is no clear evidence that these organisms metabolize lignin (18,19,20), although J. Macy (unpublished observations) has demonstrated incorporation of label into fermentation end products from the degradation of ^{14}C-cornstover lignin by a rumen fungal isolate. Further study should be carried out along these lines to establish unequivocally the role of anaerobic fungi in lignin degradation.

There is also scope for future work assessing the fiber-digesting capability of fungi in relation to diet and possible environmental implications (i.e., methane production). Aspects such as the contribution of these organisms to rumen function under different feeding regimes and their response to different manipulations and to various ionophores and metabolic inhibitors all require further investigation. Among the few but encouraging developments in this area in recent years have been investigations of the effects of metronidazole and monensin on the growth and fermentative metabolism of *Neocallimastix* (21,22). The use of various fungicidal agents and selective metabolic inhibitors should be invaluable in elucidating the role of the fungi in rumen metabolism and their interdependencies with other rumen organisms. Future work on fiber digestion should also be of sufficient breadth to include studies of novel aspects such as the little-understood role of mechanical action during the penetration of plant fiber by hyphal extension, and the possible link between mechanical pressure and the production of certain enzymes.

B. Methods for Assessing Biomass

One of the major problems in assessing the importance of anaerobic fungi in the rumen has been the difficulty of estimating their biomass. In the past, methods for estimating fungal numbers have been based on the counts of zoospores, sporangia, or thallus-forming units (23,24,25). More recently, mathematical models have been used to determine the total population of anaerobic fungi in the rumen ecosystem on the basis of the concentration of zoospores (26, Chapter 10). Another approach for the determination of fungal biomass could be the use of DNA probes. A. G. Brownlee (unpublished results) has identified unique regions of the 18S rRNA gene of *Neocallimastix*, and there is little sequence variation among the different genera of anaerobic fungi. Thus the rRNA gene would appear to be a suitable target for DNA-based probes. Further studies are being carried out in the development of this technique.

C. The Search for Anaerobic Fungi in Nonrumen Ecosystems

While obligate anaerobic fungi from domestic ruminants have been the most widely studied, these organisms are also present in the guts of feral ruminants and marsupials and in herbivores possessing major hindgut fermentations, such as the elephant and horse (27). Yet despite extensive searches for these organisms in nongut environments such as pond, river, lake, and estuarine sediments and municipal sewage, none have been isolated so far. However, facultative anaerobic fungi have been isolated from these environments and can be grown under strictly anaerobic conditions for prolonged periods (27). These organisms have been shown to possess unusual growth properties, but at present there is little information on their physiology or biochemistry. Attempts to isolate anaerobic fungi from the guts of vegetation-feeding fish have yielded interesting results. A facultatively anaerobic fungus has recently been isolated from gut contents of the mullet (28). The isolate was shown to be actively xylanolytic and to possess a powerful laminarinase. These enzyme activities possibly reflect the diet of the host, which consists mainly of green algae such as *Ulva* spp. and *Enteromorpha* spp., the cell walls of which consist mainly of xylan. Since laminarin is also a major component of many marine algae, possession by the mullet fungus of an enzyme system for its degradation is not surprising.

The search for anaerobic fungi capable of degrading algal material could lead to the discovery of new anaerobic fungus–alga associations. Indeed, the association of ascomycetes with marine algae is well known (29,30,31). These associations are either symbiotic or parasitic (29), and they may involve the obligate union of fungus to alga (mycophycobiosis), or a nonobligate lichenous association (30,31). Studies by Mountfort et al. (32) have provided the only evidence so far for an anaerobic fungus–alga association. It was observed that fragments of *Gracilaria* sp. in anaerobic roll tubes inoculated with gut contents of the mullet developed mycelial structures at their broken ends. Repeated attempts to isolate the "fungus" from the *Gracilaria* particles using a range of substrates were unsuccessful, so that continued study with the organism became increasingly difficult. One possible deduction to be made from the study is that the union of fungus with alga was a form of mycophycobiosis. Presumably growth of the fungus was obligately dependent on a set of components possessed by the alga. The continued study of these fungal forms, together with anaerobic "lichenous" fungi, should pose a considerable challenge to microbiologists in the future.

II. ENERGY GENERATION

Because they do not possess mitochondria or electron-transport components such as cytochromes and menaquinone (33,34; J. Macy, personal communication), it appears that anaerobic fungi are wholly dependent on fermentative processes to provide energy for growth. Central to the energy metabolism of these organisms are "hydrogenosomes" that are specialized for the production of H_2 (33,35). One enzyme that has a key role in this process is pyruvate:

ferredoxin oxidoreductase, which has been ascribed an important primordial function in ancient living systems (36). Less is known of the mechanisms for adenosine triphosphate (ATP) production in hydrogenosomes and whether primitive electron transport–linked phosphorylations occur involving as yet undiscovered electron-transport components. It is of interest that the ATP-generating hydrogenosomes are located near the base of the flagella in zoospores of anaerobic fungi and are connected to another group of microbodies called "kinetosomes" (33,34,35). It is well known that in aerobic chytrids mitochondria are clustered near such microbodies, presumably to increase the efficiency of energy transfer to the sites of high ATP requirement, such as for motility (37). More investigation should be focused toward understanding the biochemical basis of energy transduction between hydrogenosomes and other microbodies in anaerobic fungi, and also toward establishing unequivocally the nature of ATP production in these organisms.

III. ORIGINS OF ANAEROBIC FUNGI AND LIFE CYCLES

The role of hydrogenosomes in anaerobic fungi and other anaerobic eucaryotes (38,39) appears to be similar to that of mitochondria in aerobic eucaryotes, although the functional mechanisms for end-product and energy generation appear somewhat more primitive, incorporating enzyme systems that are ancient in origin (36). Thus there seems to be some justification for believing the hydrogenosomes may represent a premitochondrion form, and if this were so it could place anaerobic fungi at an early stage in the evolution of modern chytrids. Alternatively, if the ancesters of anaerobic fungi possessed mitochondria lost in the evolutionary path for these organisms, there is the possibility that the fungi possess a vestigial organelle that has so far gone undetected. If this is so, the accompanying DNA may still be present.

Another feature of anaerobic fungi may make these organisms unique is that their genomic DNA is remarkably rich in adenine–thymine, more so than any other eucaryote or procaryote examined to date (40). This may reflect either the compositional changes that have caused their DNA to diverge so markedly as a result of the intense competition between rumen microflora, or the ancient origin of these organisms. The life cycle of anaerobic fungi may also be more complex than originally believed. The concept of a two-stage life cycle consisting of a nonmotile vegetative phase and a motile flagellate stage appears too simplistic, and the possible existence of a sexual stage cannot be discounted, on the basis of observations of possible fusion bodies (C. G. Orpin, personal communication). Resistant bodies have also been described for anaerobic fungi (41,42). These appear to reside outside the rumen, principally in feces. The existence of these forms may add a new dimension to the life cycle of these organisms and help explain their mode of transmission and survival. Of importance will be the study of the biochemistry of these resistant bodies together with the physiological changes that occur during their formation and transition to the vegetative state, and the factors that induce these changes.

IV. NEW OPPORTUNITIES FOR BIOTECHNOLOGY AND BALANCE OF FUTURE RESEARCH

Anaerobic fungi are now being recognized for their ability to produce hydrolases effective in the degradation of plant fiber. The discovery of these organisms in other ecosystems such as marine environments (see Section I.C) will open up opportunities for the investigation of new enzymes. Among these would be those that degrade unusual algal polysaccharides, providing new products of potential value to the pharmaceutical and food industries.

Many other opportunities for potential biotechnological applications exist with these organisms. These include

1. Cloning genes of fungal enzymes. Anaerobic fungi possess powerful polysaccharide hydrolases. The cloning of the genes for these enzymes and their subsequent expression in other organisms such as bacteria may lead to improvements in areas such as waste treatment and the processing of fibrous materials.
2. The synthesis of unusual compounds. Anaerobic fungi lack the ability to synthesize some common cell-membrane constituents such as sterols because of the absence of molecular oxygen. Instead, unusual lipids and related compounds such as squalene and tetrahymenol synthesized by anaerobic pathways are incorporated into the cell membrane (43). The investigation of the ability of these organisms to produce a variety of terpenes from squalene may have commercial benefits.
3. Cultivation of anaerobic fungi for the production of bioactive compounds. To date no studies have been carried out on the capacity of anaerobic fungi to produce antibiotics and pharmaceuticals, and the field remains wide open.
4. Modification of fermentations to produce fuels or unusual metabolites. One of the few developments in this area has been the work of Mountfort and Kaspar (14), who linked the production of hydrogen by the fungi to the hydrogenation of unsaturated hydrocarbons to produce alkane fuels in the presence of palladium. There is also scope for use of the fungi in the production of volatile fatty acids for chemical feedstocks by their coculture with H_2-utilizing methanogens, or through catalytic removal of H_2.

While research into these new avenues should provide exciting challenges and opportunities in the future, it should be balanced by the continued need to better understand the role of the fungi in fiber digestion in relation to diet and possible environmental and economic implications. Although considerable progress has been made on their role in the rumen, much speculation still exists on their significance in fiber digestion and fermentation. Little definitive information exists on their contribution to rumen metabolism under different feeding regimes, although the knowledge gained in recent years has provided a sound platform for much of this work to be carried out. It is thus important that this area continue to receive satisfactory levels of funding in the future; otherwise, the benefits from the work of recent years will be lost.

V. CONCLUDING REMARKS

This chapter has detailed the author's conception of how research on anaerobic fungi might progress in the future. It draws on the knowledge gained mainly in the past decade and provides a framework for future direction of research that will lead to a clearer understanding of the role of these organisms in ecosystems such as the rumen and their possible biotechnological applications. No doubt many meaningful avenues remain to be explored that have not been detailed here, some of which may turn out to be just as important. Among these is their classification and the question whether their inclusion within the group Chytridiomycetes will stand the test of time. There seems little doubt that with new information on the genome of these organisms, the present classification may need to be reviewed. Isolation of new species and genera of anaerobic fungi (44–48) will contribute to this information and will also provide clearer ideas on the distribution, specialization, origin, and function of these organisms in anaerobic environments.

Because of their unusual properties, anaerobic fungi should provide an exciting challenge for researchers in many branches of microbiology. At present, many aspects of these organisms remain little understood, and the future should bring a wealth of knowledge about their biochemistry, ecophysiology, and molecular biology. What is remarkable is that such a group of organisms escaped the attention of microbiologists for so long.

REFERENCES

1. Alward JH, Xue GP, Simpson GO, Orpin CG. Cellobiohydrolase (CBH) from *Neocallimastix patriciarum*: a membrane associated complex? Proceedings of the XVII International Grasslands Congress. Palmerston North, New Zealand, Vol 2, 1993, p 1222–1224.
2. Calza RE. Regulation of protein and cellulase excretion in the ruminal fungus *Neocallimastix frontalis* EB 188. Curr Microbiol 1990; 21:109–115.

3. Lowe SE, Theodorou MK, Trinci APJ. Cellulases and xylanase of an anaerobic rumen fungus grown on wheat straw, wheat straw holocellulose, cellulose, and xylan. Appl Environ Microbiol 1987; 53:1216–1223.

4. Mountfort DO, Asher RA. Production and regulation of cellulase by two strains of the rumen anaerobic fungus *Neocallimastix frontalis*. Appl Environ Microbiol 1985; 49:1314–1322.

5. Mountfort DO, Asher RA. Production of xylanase by the rumen anaerobic fungus *Neocallimastix frontalis*. Appl Environ Microbiol 1989; 55:1016–1022.

6. Teunissen MJ, Hermans JMH, Huis in 't Veld Veld JHJ, Vogels GD. Purification and characterization of a complex bound and free β-1,4-endoxylanase from the culture fluid of the anaerobic fungus *Piromyces* sp. strian E2. Arch Microbiol 1993; 159:265–271.

7. Williams AG, Orpin CG. Polysaccharide-degrading enzymes formed by three species of anaerobic fungi grown on a range of carbohydrate substrates. Can J Microbiol 1987; 33:418–426.

8. Wilson CA, Wood TM. Studies on the cellulase of the rumen anaerobic fungus *Neocallimastix frontalis*, with special reference to the capacity of the enzyme to degrade crystalline cellulose. Enz Microb Technol 1992; 14:258–264.

9. Wood TM, Wilson CA, McCrae SI, Joblin KN. A highly active extracellular cellulase from the anaerobic rumen fungus *Neocallimastix frontalis*. FEMS Microbiol Lett 1986; 34:37–40.

10. Reymond P, Durand R, Hébraud M, Fèvre M. Molecular cloning of genes from the rumen anaerobic fungus *Neocallimastix frontalis* — expression during hydrolase induction. FEMS Microbiol Lett 1991; 77:107–112.

11. Xue GP, Orpin CG, Gobius KS, et al. Cloning and expression of multiple cellulase cDNAs from the anaerobic fungus *Neocallimastix frontalis* in *E. coli*. J Gen Microbiol 1992; 1448:1413–1420.

12. Xue GP, Gobius KS, Orpin CG. Isolation of a multifunctional cellulase (*cel E*) from the rumen fungus *Neocallimastix patriciarum*. Proceedings of the XVII International Grasslands Congress. Palmerston North, New Zealand, Vol 2, 1993, p 1221–1222.

13. Xue GP, Gobius KS, Orpin CG. A novel polysaccharide hydrolase cDNA (*cel D*) from *Neocallimastix patriciarum* encoding three multifunctional catalytic domains with high endoglucanase, cellobiohydrolase, and xylanase activities. J Gen Microbiol 1993; 138:2397–2403.

14. Mountfort DO, Kaspar HF. Palladium-mediated hydrogenation of unsaturated hydrocarbons with hydrogen gas released during anaerobic cellulose degradation. Appl Environ Microbiol 1986; 52:744–750.

15. Borneman WS, Hartley RD, Morrison WH, et al. Feruloyl and *p*-coumaroyl esterase from anaerobic fungi in relation to plant cell wall degradation. Appl Microbiol Biotechnol 1990; 33:345–351.

16. Borneman WS, Ljungdahl LG, Hartley RD, Akin DE. Isolation and characterization of *p*-coumaroyl esterase from the anaerobic fungus *Neocallimastix* strain MC-2. Appl Environ Microbiol 1991; 57:2337–2344.

17. Borneman WS, Ljungdahl LG, Hartley RD, Akin DE. Purification and partial characterization of two feruloyl esterases from the anaerobic fungus *Neocallimastix* strain MC-2. Appl Environ Microbiol 1992; 58:3762–3766.

18. Akin DE, Benner R. Degradation of polysaccharides and lignin by ruminal bacteria and fungi. Appl Environ Microbiol 1988; 54:1117–1125.

19. Gordon GLR, Phillips MW. Comparative fermentation properties of anaerobic fungi from the rumen. In: Nolan JV, Leng RA, Demeyer DI, eds. The Roles of Protozoa and Fungi in Ruminant Digestion. Armidale, Australia: Penambul Books, 1989:127–138.

20. Mountfort DO. The rumen anaerobic fungi. In Megúsar F, Gantar M, eds. Perspectives in Microbiol Ecology. Ljubljana, Yugoslavia: Slovene Society of Microbiology, 1986:176–179.

21. Marvin-Sikkema FD, Rees E, Kraak MN, et al. Influence of metronidazole, CO, CO_2, and methanogens on the fermentative metabolism of the anaerobic fungus *Neocallimastix* sp. strain L2. Appl Environ Microbiol 1993; 59:2678–2683.

22. Phillips MW, Gordon GLR. Fungistatic and fungicidal effects of the ionophores monensin and tetronasin on the rumen anaerobic fungus *Neocallimastix* sp. LM-1. Lett Appl Microbiol 1992; 15:116–119.

23. Joblin KN. Isolation, enumeration, and maintenance of rumen anaerobic fungi in roll tubes. Appl Environ Microbiol 1981; 42:1119–1122.

24. Theodorou MK, Gill M, King-Spooner C, Beever D. Enumeration of anaerobic chytridiomycetes as thallus-forming units: novel method for quantification of fibrolytic fungal populations from the digestive tract ecosystem. Appl Environ Microbiol 1990; 56:1073–1078.

25. Ushida K, Tanaka H, Kojima Y. A simple in situ method for estimating fungal population size in the rumen. Lett Appl Microbiol 1989; 9:109–111.

26. France J, Theodorou MK, Davis D. Use of zoospore concentrations and life-cycle parameters in determining the population of anaerobic fungi in the rumen ecosystem. J Theoret Biol 1990; 147:413–419.

27. Bauchop T. Biology of gut anaerobic fungi. BioSystems 1989; 23:53–64.

28. Mountfort DO, Rhodes L. Anaerobic growth and fermentation characteristics of a *Paecilomyces lilacinus* isolated from mullet gut. Appl Environ Microbiol 1991; 57:1963–1968.

29. Kohlmeyer J, Demoulin V. Parasitic and symbiotic fungi on marine algae. Bot Mar 1981; 24:9–18.

30. Kohlmeyer J, Hawkes MW. A suspected case of mycophycobiosis between *Mycosphaerella apophlaeae* (Ascomycetes) and *Apophlaea* spp. (Rhodophyta). J Phycol 1983; 19:257–260.

31. Kohlmeyer J, Kohlmeyer E. Is *Ascophyllum nodosum* lichenized? Bot Mar 1972; 15:109–112.

32. Mountfort DO, Asher RA, Rhodes L. Isolation of a phycomycetous-like fungus from the gut contents of mullet (abstr). New Zealand Microbiological Societies Conference, Hamilton, New Zealand, 16-19th May 1989.

33. Heath IB, Bauchop T, Skipp RA. Assignment of the rumen anaerobe *Neocallimastix frontalis* to the Spizellomycetales (Chytridiomycetales) on the basis of its polyflagellate zoospore ultrastructure. Can J Bot 1983; 61:295-307.
34. Munn EA, Orpin CG, Hall FJ. Ultrastructural studies of the free zoospore of the rumen phycomycete *Neocallimastix frontalis*. J Gen Microbiol 1981; 125: 311-323.
35. Yarlett N, Orpin CG, Munn EA, et al. Hydrogenosomes in the rumen fungus *Neocallimastix patriciarum*. Biochem J 1986; 236:729-739.
36. Kerscher L, Oesterhelt D. Pyruvate:ferredoxin oxidoreductase – new findings on an ancient enzyme. Trends Biochem Sci 1982; 7:371-374.
37. Heath IB. Ultrastructure of freshwater phycomycetes. In: Jones EBG, ed. Recent Advances in Aquatic Mycology. England, Paul Elek Press, 1976:603-650.
38. Yarlett N, Hann AC, Lloyd D, Williams AG. Hydrogenosomes in the rumen protozoan *Dasytricha ruminantium* Schuberg. Biochem J 1981; 200:265-372.
39. Yarlett N, Coleman GS, Williams AG, Lloyd D. Hydrogenosomes in known species of rumen entodiniomorph protozoa. FEMS Microbiol Lett 1984; 21:15-19.
40. Brownlee AG. Remarkably AT-rich genomic DNA from the anaerobic fungus *Neocallimastix*. Nucleic Acid Res 1989; 17:1327-1335.
41. Wubah DA, Fuller MS, Akin DE. Resistant body formation in *Neocallimastix* sp., an anaerobic fungus from the rumen of a cow. Mycologia 1991; 83:40-47.
42. Davies D, Theodorou MK, Nielson BB, Trinci APJ. Survival of anaerobic fungi in rumen digesta and faeces. Abstract No. P2-04-05. International Symposium on Microbial Ecology, Barcelona, 6-11th September 1992.
43. Kemp P, Lander D, Orpin CG. The lipids of the rumen anaerobic fungus *Piromonas communis*. J Gen Microbiol 1984; 130:27-37.
44. Breton A, Bernalier A, Dusser M, et al. *Anaeromyces mucronatus* nov. gen., nov. sp. A new strictly anaerobic rumen fungus with polycentric thallus. FEMS Microbiol Lett 1990; 70:177-182.
45. Breton A, Dusser M, Gaillard-Martinie B, et al. *Piromyces rhizinflata*, a species of strictly anaerobic fungus from feces of the Saharan ass: a morphological, metabolic and ultrastructural study. FEMS Microbiol Lett 1991; 82:1-8.
46. Ho YW. Bauchop T, Abdullah N, Jaladin S. *Ruminomyces elegans* gen. et sp. nov, a polycentric anaerobic rumen fungus from cattle. Mycotaxon 1990; 38: 397-405.
47. Li J, Heath IB, Bauchop T. *Piromyces mae* and *Piromyces dumbonica*, two new species of uniflagellate anaerobic chytridiomycete fungi from the hindgut of the horse and elephant. Can J Bot 1990; 68:1021-1033.
48. Webb J, Theodorou MK. *Neocallimastix hurleyensis*, a new species of anaerobic fungus from the ovine rumen. Can J Bot 1991; 69:1220-1224.

Index

Printed and bound by CPI Group (UK) Ltd, Croydon, CR0 4YY

23/10/2024

01778224-0012